D1453699

THE
FAMILY COW
HANDBOOK

THE
FAMILY COW
HANDBOOK
A Guide to Keeping a Milk Cow

Text by Philip Hasheider
Photography by Daniel Johnson

Voyageur Press

First published in 2011 by Voyageur Press, an imprint of MBI Publishing Company,
400 First Avenue North, Suite 300, Minneapolis, MN 55401 USA

Voyageur Press titles are also available at discounts in bulk quantity for industrial or
sales-promotional use. For details write to Special Sales Manager at MBI Publishing
Company, 400 First Avenue North, Suite 300, Minneapolis, MN 55401 USA.

To find out more about our books, visit us online at www.voyageurpress.com.

Library of Congress Cataloging-in-Publication Data

Hasheider, Philip, 1951-
  The family cow handbook : a guide to keeping a milk cow / text by Philip Hasheider ;
photography by Daniel Johnson.
      p. cm.
  Includes index.
  ISBN 978-0-7603-4067-7 (sb : alk. paper)
  1. Dairy farming. 2. Dairying. 3. Cows. I. Johnson, Daniel, 1984- II. Title.
  SF239.H37 2011
  636.2'142—dc22

                              2010040471

Edited by Danielle Ibister
Design Manager: LeAnn Kuhlmann
Designed by Pauline Molinari
Cover designed by Matthew Simmons

Printed in China

Frontispiece: Brown Swiss calf.
Title page, large: Sweeping the cow barn.
Title page, small: A Holstein in her stall.
Opposite: Rainbow over farmyard.
Contents: A Holstein and chickens.

For Mary, Marcus, and Julia

C O N T E N T S

Introduction 9

Chapter 1
Let's Get Started: Why Keep a Cow? 15

Chapter 2
Choosing Your Cow: How to Evaluate Cows 25

Chapter 3
Breed Guide: Is One Moo Better than Another? 47

Chapter 4
Housing and Fencing: Home Sweet Home. 63

Chapter 5
Feeding: What's for Dinner? 81

Chapter 6
Managing Manure: Just Compost It 97

Chapter 7
Milking: Adjusting Your Cow and
Yourself to the Process 107

Chapter 8
Calving Time: The Cycle of Life. 129

Chapter 9
Making Dairy Products: Say Cheese 147

Chapter 10
Health: Taking Care of Business. 173

Chapter 11
Calves and Kids: Raising Up the Next Generation . . 197

Acknowledgments 210
Resources. 212
Glossary. 214
Index 218
About the Author and Photographer. 224

INTRODUCTION

■ ■ ■

FRESH MILK to pour over your morning cereal, thick cream to stir in your coffee, savory butter melting on your warm toast—sounds inviting, doesn't it? Experiencing such simple pleasures on a daily basis is what owning a family milk cow is about.

Your friends may have pets, but you have something special: a family cow. Not for the bragging rights but for the down-to-earth practicality she offers. How many times have you looked at the plastic jug in your hand and thought, "There must be a better way to get fresh milk"? There is.

Start with a few acres of grassland in the country where your cow can roam. Accept the work necessary to keep her comfortable and healthy. Supply the daily labor for her care, and she will reward you with milk that can be consumed raw or handmade into numerous dairy products. Before you drift off to sleep dreaming of your partnership with a cow, however, there are important things to consider, such as selecting your animal, planning how her milk will be used, and learning the skills necessary for dairy cow husbandry. There will be challenges ahead, yet they are surmountable. In fact, they have the potential to transform the lives of you and your family.

Plan your approach to selecting a cow rather than acquire one on impulse. Identify your purpose for having a family cow before bringing one home. Establishing goals sets the stage for a more satisfying experience. For example, would you like to use your cow's fresh milk for family use only, or do you want to sell it? Do you want to convert her milk into foods such as cheese, yogurt, butter, or ice cream? Do you plan to raise other livestock, such as pigs, and feed them the excess milk? Identifying your goals and having a plan helps utilize all the milk your cow produces so none goes to waste.

This book takes you through the steps necessary to enter successfully the world of dairying. You will learn how to purchase a cow, milk her, care for her and her calf, and many other essentials in keeping her. You'll learn how to turn your cow's milk into delicious old-fashioned ice cream as well as the processes for making cheese, butter, and yogurt.

Cows are quite self-sufficient in their own right. Yet a key consideration of owning a dairy cow is the need to milk her each day and then make use of her

Opposite: Daily chores with your milk cow can be a rewarding and relaxing start and end to your day.

Keeping a cow draws everyone in the family into the picture. Caring for her increases the knowledge and abilities of each family member.

milk in some fashion. Knowing how much milk your cow will produce is useful when finding the right one to fit your lifestyle.

There is a broad spectrum in the amount of milk dairy cows give, depending on both nature (their genetic inheritance) and nurture (the intensity of their management). Due to improved management practices and genetics, many Holstein cows produce over 50,000 pounds of milk each year. That translates to an average of over 135 pounds of milk per day for a full year. At the other end of this spectrum are cows that produce as little as 4,000 pounds per year, or an average of about 11 pounds per day. Most modern dairy farms would not keep a cow in their herd that gives so little milk; it's not economically practical to cover the expenses of feeding and housing her. A low milk production cow, however, might be ideally suited to your farm and purposes. Most families would have a tough time figuring out what to do with 135 pounds of milk every single day of the year.

This handbook will be useful whether or not you have any livestock or farming experience. It starts with the basics and then develops the topics in more detail, offering suggestions and possible alternatives to the traditional way of doing things. You can adapt these ideas to fit your situation. In addition, you will learn a great deal from your day-to-day experiences and your cow's individual nuances and quirks. While there are great similarities among cows, each cow has her own individual personality and characteristics that can endear her to you and your family.

Cows on the Farm and Changes in Dairying

Having a family cow used to be more common. Even in some urban areas, many people kept a cow or two for home use in the decades prior to World War I. Agriculture began to change dramatically after the war. More science-based applications were used to solve production challenges on farms. New and improved seeds and the increased use of tractors propelled agriculture to new levels of production to meet the growing demand for food. New urban sanitation regulations soon banished the family cow to rural areas. Consequently, urban consumers began purchasing their dairy products from grocery stores.

Better breeding practices and purebred livestock were promoted as a way to improve cattle quality and a farmer's herd production. The Great Depression stymied production growth for several years with a return to self-sufficiency when many families went back to the farm. During and after World War II, dairy production again accelerated with university extension research providing information for improved animal husbandry practices.

The postwar boom created an explosion in farm production, which led to a surplus that, ironically, negatively impacted farm prices. Farmers produced too much food for their own good. While thrifty consumers benefited from this surplus, the low food prices had a profound effect on rural areas and often led to farm sales. Depressed milk prices left many farmers unable to pay their bills.

Once upon a time, every farmer kept one or two milk cows, several beef cows, a pen of pigs, some sheep, and a yard of chickens. One consequence of the fracturing rural economy was the start in specialization of raising only one species of livestock rather than several. This trend started out slowly but gained traction during the 1960s and 1970s. It accelerated during the 1980s to the point where it was a rarity, outside of certain traditional farming communities, to find farms that maintained more than one species of livestock.

Grace your family table with the rewards of owning a cow. Your cow's milk can be used to create delicious foods such as butter, yogurt, and cheeses. *Shutterstock*

Along with this specialization came housing units designed to handle large numbers of animals, whether they were dairy, beef, pigs, or sheep. This new farm dynamic quickly led to intensive management practices to increase production. Feed was brought to the animals instead of animals going out to a pasture to find their own. Machines began to replace human effort. Milking machines replaced much of the hand-milking by the 1940s. Tractors and loaders replaced the practice of hand-forking manure into spreaders. These changes were welcome and eliminated much backbreaking labor, and they also made handling a large number of livestock easier. One person could now handle more farm animals than ever before.

But the collective effect was that these changes drove people away from farms. Increased production tended to lower prices. An aging farmer population hesitated on investing in expensive equipment and buildings if they only had a few years until retirement. And, with more opportunities elsewhere with better financial prospects, young people left rural areas in droves. While it didn't happen overnight, these changes had a devastating effect for some farm families. It also created opportunities for those who could weather the storms of change.

Either a small acreage or a large expanse of land will suit your cow just fine. Your desire to do a good job will be key to your success with dairying.

What These Changes Mean to You

The changes in efficiencies of production, where more animals can be managed by fewer people, have advantages and challenges if you are looking to keep one cow. One advantage is that you are likely to locate a dairy cow whose genetic inheritance allows her to produce more than enough milk for your basic needs. Keeping a single cow, however, will likely require as much effort on your part as it did seventy or eighty years ago.

This book discusses ways to keep costs to a minimum and identifies areas where you and your family can derive great personal satisfaction, which is probably the reason you are considering owning a family cow in the first place.

A cow is a naturally sustaining phenomena in several ways. She typically produces a newborn calf each year. She converts grasses and other plants inedible by humans into milk

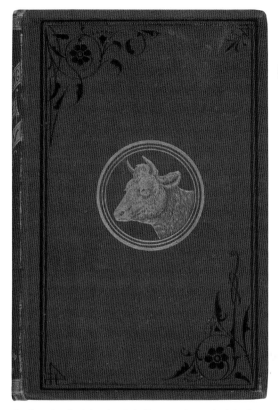

Family cow handbooks have long helped homesteaders care for their milk cow. This book, *Keeping One Cow: Being the Experience of a Number of Practical Writers, in a Clear and Condensed Form, Upon the Management of a Single Milch Cow*, was published in 1882.

and meat that can be used for human food. Her manure, a byproduct of digesting plants, provides nutrients and organic matter to the soil. This helps plants to grow and worms to enhance soil structure and is beneficial to the ecosystem of the land. By keeping a family cow, you become a valued and intricate part of a more sustainable way of life.

CHAPTER 1

Let's Get Started
Why Keep a Cow?

■ ■ ■

FARMING is about creation, whether it's milking cows and raising calves, sowing seeds and growing crops, or being a steward of the land. Many people aspire to a self-sufficient rural lifestyle where they grow their own food and live in a less-pressured setting than urban life offers. This may be your dream too. Whether you own a small acreage or are planning a move to the countryside, there are advantages to including a family cow in your dream.

Cooking with your own cow's milk makes the meals you offer your family more nutritious and delicious. You can also create a home economy with foods made from her milk. The skills you learn by working with your cow can be personally satisfying and enriching for the entire family. By owning one cow or several, you live alongside and interact with other living beings in nature. It is a power that few people have the privilege to learn and understand, let alone experience.

It doesn't take the largest acreage or the most expensive land to raise a family cow. If you have a small acreage, she will do well so long as there is good soil that produces an abundant crop and has shelter available; or if you have a larger piece of land, she will be able to thrive on rough, rock-strewn soils if there are sufficient grasses. Keep in mind that with only a small acreage available, you will need to buy extra hay to meet her needs during much of the remaining year when grasses are dormant. One acre won't produce enough grass for her to graze and from which to make winter feed. Also, your cow needs daily exercise so a minimum of one to two acres is recommended. Your couple of acres of grassland can be processed through your cow and converted to food for your family. You can have a hand in making a vast array of products from her milk. And your family will witness the miracle of nature's cycle as you progress through each year with your cow and her annual calf.

Opposite: Petite or large, your newly acquired cow heralds new experiences for you and your family. You're about to embark upon a great adventure with many opportunities for learning new skills.

It's a pleasure to own a docile, friendly dairy cow. As you provide daily care for her, she soon becomes part of the family.

What's Your Plan?

Before bringing a cow home, make sure you live in an area that allows livestock to be raised there. Most, if not all, towns, villages, and cities have ordinances that restrict animal agriculture within their boundaries. Therefore, a country residence is generally the first step in cow ownership, followed by being willing to commit your time and effort to provide your cow with all the comforts she'll require.

There are some practical dos and don'ts of bovine ownership. Equipping yourself with information and having a well-thought-out plan saves you from some headaches and disappointments later.

Start your plan with a simple exercise. List all the reasons you have for owning a cow. (Sustainable lifestyle? Good for the kids? Money saver?) Next, put down your goals. (Constant supply of fresh milk? Make cheese? 4-H projects for kids?) Don't worry if you don't know a thing about a dairy cow or what you would do with her. This book will help you learn about cow care and provide knowledge of the steps to take to be better prepared.

Do plan to devote time each day to care of your cow. This care includes milking her; providing feed, water, and a clean place to live; and observing for signs of discomfort or illness. You cannot justify the expense of owning a cow if you are simply trying to save money on food. There has to be something more compelling that drives your desire to invest in the capital costs involved and the time commitment required. There will be a partnership between your family and your cow. In return for the food, water, and shelter you give her, she will give you milk for drinking or processing, manure for composting, and a calf every year that can be raised or sold.

If you expect your cow to provide a source of income, look beyond selling fluid milk (which may not even be permissible in your area). Consider value-added products, such as cheese, butter, or yogurt, that can be processed for sale at farmers' markets. Many people are already doing this, and you can too.

The satisfaction value of owning and caring for a cow can exceed the costs of maintaining her. Starting and ending your day with her can be the pleasant and rewarding experience to compensate you for your investment and hard work.

Calculating How Much Pasture Your Cow Needs

The amount of pasture or paddock area your milk cow needs depends on a few factors. One is her size: a bigger cow needs more food than a smaller cow. Another factor is the length of the growing season in your region. If you live in California, your cow may be able to graze outside year round. If you live in Wisconsin, your cow only has about six months of pasture, and the rest of the time you need to provide her with hay that you purchase or grass that you harvest from your own field.

Typically a mature dairy cow requires about 4 percent of her body weight in food. Thus a 1,100-pound cow requires about 30 pounds of forage dry matter (DM) each day. Well-grown pastures of moderate density or thickness will yield between 2,000 to 2,500 pounds of total dry matter per acre (DM/A) in the first 8 inches of growth. You can use the following formula to determine the amount of paddock or pasture area needed based on her body weight requirements:

(Number of animals × DM intake × days in paddock) ÷ lbs. forage/acre = pasture size

As an example:

(1 cow × 30 lbs. DM × 180 days) ÷ 2,250 lbs./acre = 2.4 acres

This is the total acreage needed to produce enough hay and grass to feed her during the pasture season. One cow is unlikely to graze this efficiently due to rapid plant growth in spring and early summer. If this amount of acreage is available, you may be able to cut, dry, and bale the excess and store it for winter feed. In most northern climates, two to three cuttings can be taken off the fields during the growing season.

From rock-strewn fields to pleasant prairies, a variety of landscapes are appropriate for a cow, providing she has sufficient grasses to eat.

Consider Your Options

A cow can consume a lot of grasses and other plants each day. How will you use your land to feed your cow? Will you supplement her diet by purchasing hay? These are questions to answer before purchasing a cow.

Stocking rate is a calculation for determining how many animals can be supported by certain acreage. This rate is easy to calculate. Use it to determine how much hay, if any, you may need to purchase during the year. Your location and climate determines the length of the growing season. This impacts how much grass is available for grazing and what needs to be harvested for winter feed.

Since cows are ruminants, they make great use of any existing grassy areas. They keep many areas around the farm clipped down, although they generally won't eat certain weeds, thistles, or brush. Many farms have places not easily accessible with machinery, such as steep hillsides or rocky areas. A cow can reach these spaces with little trouble and utilize them for feed. If you have good pastures, your cow could also be a companion animal for sheep or goats.

Your cow will produce an abundance of manure. While she grazes, she returns this organic matter and its nutrients back to the soil. In a barn or shed, manure can be hauled out for compost and used later in your garden.

If you set your sights on a simple initial goal, such as having milk for your family table, you allow yourself time to learn basic cow care before developing larger-scale alternative uses for any extra milk. Your options may expand or change as time goes on, but it begins with a single cow. You may later consider having more than one cow. It all depends. How much milk do you need? How much time can you commit? How much land do you have?

Cheese-making is one way to make use of extra milk. You can make cheese from any milk, be it cow, goat, sheep, or even water buffalo. Species differ in quantity of milk produced daily and butterfat and protein content. Yet the basic process of converting their milk into cheese is virtually identical.

One advantage of cheese-making is that it ages over varying time frames. Even when your cow is in her unproductive period before she has her next calf, you can enjoy the fruits of your labor in a wedge of homemade cheese.

Extra milk and whey, a byproduct of cheese-making, can also be used as feed for other livestock, such as pigs.

Cows are social beings often comfortable around other species, such as sheep and goats. This makes them good companion animals and allows you to keep one cow without having to find a suitable partner cow. However, a cow can still do well even if she is the only animal on the farm.

Cows are social creatures. Companion animals can include goats, sheep, llamas, horses, or any other animals that have social tendencies.

A dairy cow, even with modest production, greatly surpasses other milk-producing livestock, such as a goat or sheep. Taking into consideration all dairy breeds in the United States, a cow produces an average 20,500 pounds of milk each year. This equals about 56 pounds per day, or approximately 7 gallons. Of course you may choose a breed that produces as little as 2 gallons per day. This amount may not seem like too much initially, but you must plan on doing something with that much milk every day, seven days a week.

Taste that fresh milk. Your cow's milk can be made into a wealth of dairy products, from butter to cheese to yogurt or ice cream.

Abundant milk production leads to several possibilities. The excess above what is used in your household can be sold, depending on state laws regulating the sale of raw milk. It can be given away, or it can be processed into a variety of dairy products, including yogurt, ice cream, and cheese. If you own other livestock such as sheep, their milk can be mixed with cow milk to produce an even wider variety of tasty and attractive cheeses.

Involve Your Family

Care of a cow offers a set of responsibilities for younger family members. They'll learn how to handle and manage livestock. Owning animals requires commitment to their well-being. Committing to a daily schedule or routine helps develop time management skills. While certain risks are always present with large animals, younger family members can take steps to stay safe and along the way learn useful skills that may lead them to careers in animal-related or scientific fields.

Cattle have their own life rhythm in which they grow. The land and nature also have a rhythm of their own. Likewise, you will develop a rhythm with your cow as you set up a daily routine. The chores that are part of working with a cow can become a quiet, almost meditative time rather than be regarded as work.

A cow requires your time each day to attend to her needs and to milk her. But the rewards you receive from your daily efforts can be very satisfying and fulfilling. In the end, that is a worthy achievement.

Morning chores usher in the day as a peaceful time to commune with your cow.

Help Along the Way

This book answers many questions you have about the fundamentals of getting started with your family cow. As you gain experience in handling her, you will have even more questions that require in-depth answers. Those answers may exceed the scope of this book. There are many resources for help and advice.

Many counties have an **agricultural extension service** where trained educators can access information from state universities and research departments and pass it along to you. Extension personnel have a network of contacts across the country and, increasingly, around the world. Although you may have the same access to relevant agriculture data on the Internet, their background and experience can help interpret this data, and their services are free.

Technical schools and colleges offer classes that may be of benefit to you as a beginning dairy farmer. You might take a class that offers training in milk processing, such as making cheese and butter, or in business development.

Grazing networks are an excellent way to get to know other farmers in your area who use grass-based feeding programs. These informal groups can help beginners solve problems, explore alternative methods, gain social contacts, and have fun learning from the discussions. Grazing networks are typically a mix of livestock farmers who have experience in dairy, beef, sheep, or goats ranging from experienced to novice. Often their discussions reflect helping others rather than imposing views upon new members. Many of these networks use a series of pasture walks to observe the management scheme a host farm is using. You may even find yourself hosting one someday. Most agriculture extension offices are aware of the grazing networks that exist in their county. Don't be afraid to ask about how to attend. They can be great fun for the whole family.

Owning a farm and keeping a cow can introduce you to a rewarding lifestyle. Consider the possibilities, develop a plan, and use your imagination and efforts to bring it forth.

All **land-grant colleges** in the United States provide agriculture classes that are available to the public. The university system can provide research information on a variety of subjects. These reports may help in decision-making related to your farm or family cow.

The **United States Department of Agriculture (USDA)** is the cabinet level agency that oversees the vast national agricultural sector. Its duties range from research and food safety to land stewardship and resource management. Every state in the country has a **state department of agriculture** that administers the programs of that particular state and operates under its statutes. These agencies have a wide range of booklets, pamphlets, and other publications available to help you understand the rules and laws of livestock ownership and food sales, how to obtain licenses, and how to comply with regulations pertaining to your farm.

If you farm and have a family cow, you will come into contact with the USDA in one form or another. That's good because there may be several farm programs available to help you get started and manage your property in an environmentally sound manner.

The **Internet** has a lot of information that may help you. However, check that the sources are well-informed and knowledgeable and be sure to protect your personal and farm data if responding to outside requests.

Opposite: Consider joining a grazing network. You'll gain ideas about how to use the pastures you have available. It's also a good way to connect with like-minded farmers.

23

CHAPTER 2

Choosing Your Cow
How to Evaluate Cows

■ ■ ■

SELECTING your family cow is like buying a family car. It can be thrilling. It can be nerve-racking. It can mean many years of enjoyment. Choosing wisely is the key. Much like a car, you want a cow that fits your needs: fills your pail (think good gas mileage), is healthy (in good working order), and is a pleasure to work with (luxury sedan). Where do you find such a cow?

There are several good public and private options: a neighboring dairy farm, a dairy cattle auction barn, a dairy cattle dealer, a farm sale, or an advertisement in a dairy farm publication. Ask for help if you are not sure of your own judgment. Help may be available from another dairy producer, a county agriculture extension agent, or someone else who has dairy cattle experience that you trust. While many people involved with agriculture are honest and well-intentioned, the caution of buyer beware is still a good policy to follow. Don't be afraid to ask questions. Listen closely for straightforward answers. When things don't quite seem right, look elsewhere.

Check with Neighbors

If you live in a dairy-producing region, your cow may be as close as a neighboring farm. Dairy farmers cull cows that have low production from their herd and send them to a livestock market. There may be nothing physically wrong with the cows; it's only because their production was not high enough compared to their herdmates. Remember that your

Opposite: Consider the animal you want to start with. A calf grows into a productive cow with proper nutrition. This may be a good way to start.

This Cow Milks!

The world milk record is held by a Holstein cow. She produced over 76,000 pounds of milk in 365 days. She averaged over 208 pounds per day, every day, for a full year. This is the extreme production end, obviously, for milk cows. The cow you buy simply needs to be physically sound with adequate milk for your needs and priced at a level you can afford. You don't need a Rolls-Royce when an economy model will do.

needs are different from the large-scale dairy farmer. A young, healthy cow with low milk production can be an excellent buy for you. She won't cost as much as others, and besides, you don't need the farmer's best-producing cow for home milk production.

Once at the livestock market, three things may happen to culled cows. A cattle dealer may purchase them for resale to other farmers. A meatpacking house may buy them for slaughter. Or, they may be purchased by other dairy farmers who find the cows productive enough to add to their herds.

One advantage of buying a cow from a neighboring farm is that the farmer's reputation is at stake in whatever animal he or she sells to other producers. This can work in your favor. The farmer is likely to be upfront about the true nature of a cow's health and milking ability. Word travels fast in a rural community. Selling a misrepresented cow to someone who lives nearby would backfire in the long run. Neighbors have a vested interest in their future business by answering your questions honestly.

There may be several reasons a farmer is selling a cow other than low production. She might not be able to become pregnant. Or, she may have a physical problem, such as lameness or other mobility issues. Digestive, respiratory, or intestinal problems and mammary system infections may not be easily recognized just by looking at a cow.

Two diseases, tuberculosis (TB) and brucellosis (Bang's), are controlled by state statute, although farmers are not required to vaccinate against either. The diseases can surface in a herd because of exposure to wild animals, such as deer.

If tuberculosis is detected, the cow herd must be quarantined and testing must be done to identify any disease carriers. Carriers, when found, are immediately destroyed.

Brucellosis, otherwise known as Bang's disease, is a reproductive disease that causes abortion in a cow and reproductive problems

Those Who Can Help You

- Dairy farmer
- County ag extension agent
- Person experienced with dairy cattle
- Dairy cattle sale manager

Places to Buy a Cow

- Neighboring dairy farm
- Dairy cattle auction barn
- Dairy cattle dealer
- Farm sale
- Advertisement in a dairy farm publication

A neighboring dairy farm may be a good place to find a family cow.

later on. It can cause a disease in humans called undulant fever, which is transmitted by drinking unpasteurized (raw) milk from an infected animal.

Many states have an eradication-free status for these two diseases, and problems only arise when cattle from other outside states are brought into a herd. You should ask about the health status of the herd and the cow you are considering—that is, whether the cow has been vaccinated for either disease. You might even ask the owner if you can consult his or her veterinarian to perform the tests to ensure they are negative for both. If the owner is hesitant, offer to pay for the vet. The tests are small insurance when you consider that buying a positive animal and then having to sell it later is a dead-end proposition. If the owner still hesitates, look for a cow at another farm. (Be aware that the farmer may hesitate because he or she just doesn't want to know themselves, and having the cow test positive for either TB or Bang's would trigger a quarantine and thorough testing of the entire herd.)

A veterinarian can also run other tests, such as the somatic cell count of the milk (if the cow is milking), which indicates mammary disease, and a pregnancy check to make sure the cow is carrying a calf. If taking a milk sample, the vet can also examine the cow's teat structures. The teats should be open and functioning correctly with no obstructions or damage from injuries, which would make milking her more difficult. The vet can also tell you whether all four quarters of the cow's udder are functioning and if there are any hard spots in the tissue indicative of a previous infection or injury.

If you do come across a cow that milks on three quarters but otherwise is in good health, don't necessarily discount her. Remember that your personal criteria for a useful family cow are different than those of a large dairy herd. She might be a cheaper way to get a great cow for your needs.

Look at Auction Barn Sales

In some areas, auction barns specialize in dairy cattle sales at different times of the year. These sales can be biweekly, monthly, or at other scheduled times. Sometimes an auction barn hosts a dispersal sale, wherein all the cattle are transported from a farm to the auction barn to be sold.

A dispersal sale is a good way to buy your cow since the reasons for the sale may be totally unrelated to the productive capacity of the cows. The dispersion and sale of an entire herd may be offered due to the farm owner's health, a death in the family, retirement, or a host of other reasons.

During the sale, the auctioneer may comment on the health status of the herd and point out particular cows he feels are the best animals. Since he has been hired by the farmer, the auctioneer has a vested interest in the cattle bringing the highest prices possible. This is a legitimate approach, but you need to use your own caution in assessing the worth of an animal. Sales pitches and the lingo that goes with the auction process can be a distraction.

Reasons Cows Are Sold

- **Economic**—they don't produce enough milk.
- **Health**—seen and unseen physical problems or disease.
- **Situational**—more cattle than barn space allows, farmer no longer dairying.

Health Tests

Having information about your potential purchase is better than being in the dark and taking home a cow that may cause you problems from the start. Seek a veterinarian's assistance to run tests and give your cow a clean bill of health.

Generally, cows in the last half of their lactation, or near the end of the sale, sell for less than cows that are soon-to-calve or recently fresh and producing lots of milk. The highest bidder gets the cow and pays for her before loading her on a trailer to go home. If you need to arrange for transporting your cow, the auction manager may be able to recommend someone.

An advantage of auction barn or farm sales is that the cows offered vary in stages of lactation. Many dairy farms try to maintain an even number of calvings each month in order to have a consistent milk level flow throughout the year. Herd dispersals offer some pregnant cows that will calve in a short time, some not yet bred, while others may be halfway through their lactation and not ready to calve for another four or five months.

This may fit your situation very well, particularly if you're not quite ready to begin milking. You can bring a cow home and allow her to become accustomed to you and your farm.

Dairy cattle can be purchased at farm auctions, such as the one shown here. Auctions are held to sell the farm assets, which usually include cattle.

Make your family part of the selection process. It will give them a sense of ownership in the cow you bring home.

Animals sold at herd dispersals are sold "as is" unless other announcements are made at the time she sells. You have little recourse for any reimbursement if something goes wrong after you get your cow home, even if you think something was overlooked at the sale. Still, you can always discuss such problems with the auctioneer later to see if some accommodation can be made to resolve any problems. Most auctioneers try to keep their farmer clients and their buyers happy. If something does goes wrong, don't be afraid to ask.

Consignment sales are much like a herd sale in terms of the auction setting. Farmers consign a top animal from their herd, hoping to secure a premium price. The consigner pays a commission fee to the auction management; however, this does not affect the selling price.

Fresh Cow

A cow cannot milk uninterrupted forever. She must freshen her milk supply regularly by calving, which induces her to begin producing lots of milk again. Thus a "fresh" cow is one that has recently calved.

Above and below: Some farms have converted their business into a dairy cattle auction barn. Sales may be held on a regular schedule.

It's worthwhile to visit several auctions or herd sales before making any purchase. By doing this, you'll glean an idea of what cows cost based on different stages of lactation, age, size, and perhaps production levels. This information is useful, too, if you decide to buy directly from a farmer or cattle dealer.

Low-priced cows are not necessarily a bargain. If an auction barn does not have any health, pregnancy, or production records available to examine, you have little information on which to base a purchase. The cow may have unknown health issues. She may not be pregnant or milking well at the time of the sale. That means you'll wait a longer time for a calf and for her to start milking. However, if you don't need a lot of milk right away or don't want a calf to care for very soon, then she may be an excellent buy for your situation. Often this kind of cow sells for just above beef market price. Should she not be pregnant or produce very much milk, you can then send her to market and recover most or all of your investment and go buy another cow.

Advantages of Buying at Auction Sales	Types of Auction Barn Sales
• Lots of choice in cows • Cows in various stages of lactation • Health tests may be available • Trucking may be arranged	• Regularly scheduled sales • Herd dispersals • Consignment sales

Cattle Dealers Offer a Service

A cattle dealer buys and sells animals to make a profit. His or her best interests may come before those of the seller or buyers, so be more cautious in your purchase. There are many reputable dairy cattle dealers, however. Farmers in your area often know who they would trust when buying a dairy cow. Ask them to recommend a cattle dealer they feel is trustworthy, and work with one of them. An advantage of using a cattle dealer is that you may be able to have your cow delivered to your farm rather than having to arrange transport yourself. Most cattle dealers have trailers that make loading and unloading cattle easy and safe.

If you aren't sure how to evaluate a cow on your own, enlist someone to assist you, such as a neighboring farmer, a large-animal veterinarian, or a county ag agent.

Watch Advertisements for Cattle Sales

Cattle for sale are featured in farm newspapers. Some of these papers are weekly publications and contain advertisements from dairy farmers, dealers, and others who are selling a few head or many. Auction barn sales and farm auction announcements can also be found in these publications.

Never buy a cow sight unseen. You can call and get details about the cattle advertised, but check for yourself in person that it's a cow you want bring to home.

Need Some Help?

Not sure about your own judgment in selecting a cow? There are many knowledgeable people you can ask for help, including:

- **Other dairy farmers**
- **County agricultural extension agents**
- **Large-animal veterinarians**
- **Breed association representatives**

Two sets of eyes are better than one. If you think you need some help in selecting a cow, don't be afraid to ask.

Low-clearance trailers make loading and unloading cattle easy. Your cow can be transported to your farm in a trailer such as this one. Many custom-hire haulers use this type of vehicle and trailer.

What to Look For

Your cow needs to be more than just a pretty face. She needs to be useful and functional, and this involves assessing her physical qualities. Do you want a young cow that will grow old on your farm? Do you want an older, proven cow to start with and then replace her in a few years with a younger one? You'll want to learn what physical attributes to look for in a cow, how to assess her disposition, and how to evaluate her body condition and health. If all these are satisfactory, her age will be less of an issue.

Looking at a Dairy Cow

These are some traits your cow should possess:

- **An angular, feminine appearance**
- **Alertness with a calm disposition**
- **Thin, pliable skin and glossy hair-coat**
- **A straight topline, smooth-blending shoulders**
- **A wide, square rump**
- **A well-attached high udder, soft and pliable**
- **Squarely placed teats**
- **Good body capacity**
- **Sound feet and legs**
- **A deep heel**
- **Ease in walking**

The Parts of a Cow

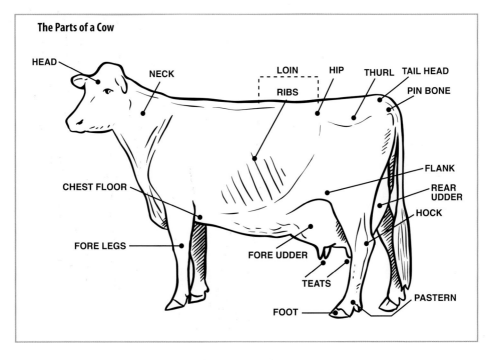

Conformation

A cow's attractiveness is related to her general appearance. This is influenced by her physical conformation, or how her body parts are put together. Your cow doesn't need to be a prize-winning show animal, but there are certain shapes, lines, and skeletal formations that lead to attractive and healthy, durable animals.

When assessing a dairy cow, think in terms of triangles. A good milk cow approximates an image of a triangle, or wedge shape, when viewed from three different angles: from the side, from the front, and from the top of her back from the rear end forward.

These triangle shapes indicate an angularity associated with a dairy type of cow. Depending on the stage of lactation, some cows may have a "skinny" look. Heavy milk production burns off body condition since a cow is eating to feed herself plus make milk. As a cow progresses in her lactation and milk production starts to drop, she begins to regain some of the body condition she lost earlier.

The Dairy Cow Triangles

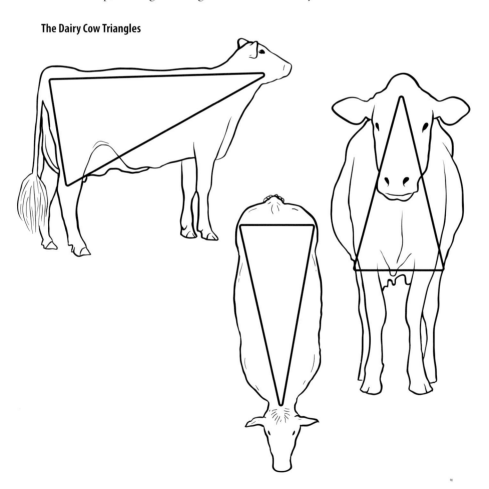

Above: Thin body condition often comes from high production during lactation. In a dry cow, it may mean she didn't receive adequate nutrition.

Right: This is a cow with heavy body condition. Excess fat can cause health problems if not managed correctly following calving.

She'll look a bit thicker than she did earlier. Even with added body condition, these triangles remain apparent and are an indication of milking ability.

Cows that don't exhibit any wedge shape are also the types that don't milk well for an entire year. These cows tend to look "beefy" and are more likely to turn their feed into meat rather than milk. Depending on how much milk you want her to give, a non-dairy-looking cow may be suitable for you.

A cow should have a wide muzzle and open nostrils, clear eyes, and shoulder blades that smoothly attach to her body wall. She should have a feminine head; a long, lean neck; and a straight or fairly level spinal column, or topline, from her pin bones and tail setting to the shoulder as viewed from her side. Her front legs should be set apart and not bowed, crossed, or too close together at the feet. Being wide apart at the feet indicates a good chest capacity that allows room for her heart and lungs to work properly. Good body width allows for more room to comfortably house internal organs and stomachs.

A feminine, attractive head typically correlates to the rest of the cow being the same way. It is a sign of good dairy character or milking ability. A long, lean neck; pliable skin; and soft hair are also general indicators of a good milk cow.

This young cow shows width between her front legs. This width indicates body capacity that allows room for her heart and lungs to function properly. Note the sturdiness of her front legs as well.

A cow that has a short neck, heavy shoulders, and thick rear quarters looks more like a beef cow than a dairy cow regardless of her color. She also lacks length in her ribs, which indicates more strength than dairy qualities. She may not be as high producing as other cows but still may suit your milk needs.

Try to buy a cow with a straight topline. A low back on a cow, as pictured here, does not necessarily signify any physical problems; however, she typically has a deep belly and may be lower set to the ground. This is not a problem unless she also has a deep udder, making it more difficult to milk to her.

These cows possess width of rear end, indicated by wide pin bones, thurls, and hips. This is important because it makes it easy for a calf to be born and allows room for the udder to suspend properly.

Right: The open hock structures seen on these young cows are flexible, which gives the cow an easier motion while walking. When a cow's legs are too straight, they do not flex as readily, which can lead to leg or udder injuries.

Her hind feet should have heels at least 2 inches deep. This minimizes foot injuries. A flexible joint in the rear leg called the hock aids in mobility and reduces leg stress. Hind legs that are too straight will cause feet problems, similar to you standing on your tip-toes all day long. That kind of cow will lie down a lot during the day because her feet hurt. Another extreme to avoid is a cow with hind legs that are too open in the hock joint. These look "sickled," like a jack knife that is half folded shut. While a cow with this type of leg moves easily, she usually has a shallow heel that is more susceptible to foot injuries because it has a lower foot angle touching the ground.

A cow's udder should be symmetrically shaped with four teats set equally apart of similar size and, if you plan to hand-milk, that are long enough to grasp with four fingers and the palm of your hand. Teats can be shorter if you are going to use a milking machine. A cow's udder should be positioned high into her

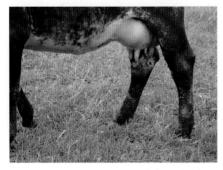

Long teat structures, such as those pictured here, are ideal for hand-milking. *Marcus Hasheider*

Short teat structures, such as those pictured here, are desirable for milking machines. There is less tissue to put into the teat cup.

A pendulous udder is one that extends below the hocks. Avoid buying a cow that looks like this. They are more prone to injury and mastitis because their udder is closer to the ground.

This is an example of a strong median suspensory ligament. It divides the udder into two halves and keeps the udder up and off the ground, away from injury.

pelvic cavity and not extend below her hocks in depth. Avoid cows of any age with swollen, pendulous udders that are close to the ground. You will have trouble getting a milk pail under this type of udder for hand-milking or even getting a milk machine to do an adequate job.

Cows with large udders are sometimes thought to give more milk because there is more room for milk than in a small udder. That is rarely true. Usually a large, pendulous udder signifies major damage to the udder tissue or a torn suspensory ligament. Avoid buying a cow like this at any price. Large, pendulous udders are also prone to infections because their teats are near the ground and can pick up dirt, mud, and manure. Bacteria grow through the teat ends, causing an infection called mastitis, which, if severe enough, can cause death. Cows with large, pendulous udders are simply not worth the trouble they cause you.

A good udder is one that has length. It provides even spacing of the teats between the left and right halves of the udder and between the front and rear quarters.

39

There are many indicators of a healthy cow. Clear, bright eyes are one of them. The more cows you look at before you buy one, the better job you will do in your final selection.

Watch a cow walk and turn around before you buy her. Look for an easy gait that does not show signs of lameness, limping, or sore feet. A cow that has flexible, durable legs and joints is less likely to develop problems quickly.

Look at her eyes. They should be clear and bright, not lackluster or dull. Steer clear of a cow that has discharge from her eyes or nose or labored breathing. She should be responsive to what's going on around her but not to the point where she'd rather jump the fence than look at you. Avoid any cow that appears to be nervous, agitated, or mean. These conditions may have been caused by her handling rather than her nature, and some cows may be tamed with patience and diligence to bring them back to a more passive state. But life's too short for an extended hospital stay because a cow mistook you for a kicking board.

Another subtle characteristic of a good dairy cow is her hair. It should be glossy and smooth rather than coarse when you feel it. Her skin should be thin, pliable, and elastic as you pull it from her body.

Her pin bones, thurls, and hips should be wide apart and level. The pin bones are the points at the rear of the cow protruding out from the pelvis. The thurls are the hip joints where the rear legs attach on either side of the cow's pelvis. Width in these areas allows for greater ease in giving birth and for adequate room to house the udder underneath the pelvis. Very narrow rear pins, hips, and thurls typically create constrictions in the birth canal, making it difficult for the calf to exit. It also places pressure on the top portion of the udder. This tends to force the udder downward as it expands at calving time or during high production. This downward pressure stresses the median suspensory ligament, which acts like a rope and holds the udder up into the body cavity. If this ligament stretches too far or breaks altogether, there is nothing left to hold the udder up off the ground. The udder tissue can hold it somewhat, but with the loss of the ligament's support, the bottom of the udder begins to fall away from the body, and the udder tissue gradually stretches and tears, causing bleeding and edema. This greatly

A narrow rear end, such as pictured here, does not provide as much room for the calf to be born as an open rear-ended cow.

enlarges the whole udder—the very pendulous udder you're trying to avoid buying in the first place.

A cow should have a soft, pliable udder with no hardness or scar tissue in any of the four quarters. Scarring or lumps in an udder are signs of a previous infection. This reduces the amount of milk she will produce in subsequent lactations. Make sure all four teats work properly. There should be no obstructions at the ends or in teat canals. If a cow is milking, go ahead and do a hand-milk test run. Strip her out in each quarter, and make sure her milk is not clotted or red-colored, which would also indicate unhealthy udder tissue. Your money is best spent on a cow that has four working cylinders.

Most cow breeds have horns. While they may be attractive, they can also be dangerous. She can use them as weapons if she wants to. It is easier to remove horns on young animals than to do it when they are older.

The polled gene is dominant and does not allow the growth of horns. This trait can be incorporated into breeding programs, thus eliminating the need to dehorn animals.

No Horns

Horns may be cute, but they can be deadly. Most dairy cows have their horns removed when they are young or are naturally "polled"—a genetic trait that can be incorporated into a breeding program. Horns can cause injury to handlers even when no malice is meant from the cow. An excited cow, especially after giving birth to a calf, can be a very protective mother. She may swing her head to try to ward off any intruders she deems a threat to her calf, even her owners whom she's seen every day. If the choice is equal between two cows, one with horns and one without, it's safer for you and your family to choose the hornless cow.

A first-calf heifer is a young cow that has had only one offspring. She typically has a smaller udder than older cows. If she is sound physically, she has the potential to last for many years of milk production.

Appropriate Age

Young cows are generally healthier than older cows. They also have several years of production ahead of them. Yet older cows usually have proven themselves to possess many good qualities; otherwise, they would have been culled from the herd years ago. Most modern dairy farmers do not tolerate cows with physical abnormalities, breeding issues, or health problems. A cow that is five to ten years old has avoided being culled for any of those reasons.

Consider your own goals. If you want to make a long-term commitment to milking, then a young cow that has the potential to stay around seven or eight years may be your best choice. If you are only considering a one or two year milking plan, an eight- or nine-year-old cow may be the right choice. Don't discount the potential of a cow because of her age alone. An older cow may have several useful years left, particularly if relieved of the stress of optimal milk production in a large working dairy. While age considerations are important, a cow's physical conformation has the most impact on whether or not she can stay around for many years.

Pedigreed or Not?

Purebred, or registered, cattle tend to bring higher prices than grade, or unregistered, ones. A registration paper alone does not guarantee a better cow than her unregistered counterpart. A pedigree may not be an important concern for a family cow. However, if you want your children to be able to show your heifer calf at breed shows and state fairs, the cow usually needs to be registered in her breed association. Only a handful of shows, such as county fairs, may be open to unregistered cattle.

Don't discount older cows just because of their age. Just make sure they are healthy animals with no udder and feet problems.

Stage of Lactation

The stage of lactation refers to the milk-producing period between calving and drying off from milk production before the next calf, such as early lactation, mid-lactation, or late lactation. A dry period is when a cow is pregnant and not producing milk. Her body is resting before she has her next calf. There are advantages to buying a cow during her dry period. She has time to adjust to her new surroundings and feeding program without a decrease in milk production, and you will be presented with a calf in a short time.

Cows purchased in the late stage of lactation, shortly before drying off, are often cheaper. Buying a fresh cow that recently had a calf means you need to milk her as soon as she arrives at your farm. In that case, you need to prepare for her arrival ahead of time and have a plan for milking her.

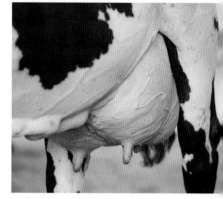

Recently fresh cows will have an extended udder, as pictured here. This results from a combination of edema and milk production at the time of calving. Most cows quickly remove the edema and swelling through the blood supplying their udder.

Other Considerations

There are other physical conditions that you should look for but may not be as readily apparent at the time of purchase. Try to observe any vaginal discharges, urinating difficulties, and the manure consistency before bringing her home.

A healthy vaginal discharge looks like a clear string of mucous, which indicates a healthy uterus and reproductive system. These discharges often appear during an estrus cycle, or heat period, when she's ready to be bred. A bloody discharge appears as the heat period ends. Neither of these should be in profuse amounts, and both are normal. A whitish or cloudy discharge indicates a uterine or vaginal infection, which can usually be cleared with appropriate therapeutic treatments. In severe cases, these infections can render your cow sterile or a problem breeder.

Your cow's urine should have a clear, yellowish color, which typically indicates healthy kidneys and bladder. Any blood in the urine could indicate an infection.

Depending on the feed available, a cow's manure may have different consistencies. Spring grass causes a loose, green manure, while a diet of dry hay in winter makes the manure more fibrous and firm. Watery manure is never normal and usually indicates a problem. If possible, try to observe a cow's manure and method of defecation before buying her. She will normally arch her back while pushing the feces out. Standing level-backed and defecating is not normal and could indicate that she may have swallowed loose wire or other hardware that is now causing problems in her stomach. Also, she should lift her tail while defecating. Not doing this usually indicates nerve damage in her pelvic area, which may be the result of a difficult birth or mishandling by workers. This condition may lead to breeding problems because the fecal material seeps back into the vaginal recesses, causing low-grade infections and poor reproductive performance.

There is one major disease that you should be aware of and often can be detected by visual observations but more conclusively by a serum test: Johne's disease (see page 177). Cows in an advanced stage of this disease have watery, pea-soup manure but appear to have a good appetite. Do not buy a cow that shows these signs, no matter how good she may look to you.

There are many things to consider when buying a cow. Take some time to acquaint yourself with cattle by attending auctions and sales before you buy your cow. You will find these enjoyable social experiences. You will also learn tips about cows to make better comparisons. It won't be long before you begin spotting the good ones.

Observe a cow before buying her. A normal posture to defecate or urinate involves arching her back. If she does not do this there may be a problem.

A cow that is not milking and is dry, or not lactating, has an udder that collapses to almost nothing. It is a normal situation and a dry udder should look like the one shown here.

Breed Guide
Is One Moo Better than Another?

■　■　■

EACH BREED has a distinctive look, be it color, size, or other more subtle characteristics. Cows come in a rainbow of colors. Do you like brown cows? Orange cows? Black and whites, or reds? There are twelve breeds designated as dairy cattle in the United States. Each has a distinctive color that will help you identify them. Among these twelve breeds, there are several that are considered very rare.

The major breeds include Ayrshire, Brown Swiss, Guernsey, Holstein, Jersey, Milking Shorthorn, and Red and Whites. Other breeds not as prevalent include American Lineback, Dexter, Dutch Belted, Milking Devon, and Normande. There are also crossbreds, which result from mating one breed with another for one generation and then to another breed in the next.

Most breeds of cattle adapt to any climate as long as there is access to feed, water, and shelter. For your purposes, your region's climate will have more of an impact on feed production than on the cattle themselves.

A major difference between breeds is in the average components of their milk. Some breeds, such as Jerseys and Guernseys, have higher percentages of butterfat and protein in their milk, on average, than Holsteins or Milking Shorthorns.

Milk with higher components percentages produces a higher yield of cheese and butter from the same quantity of milk than milk with lower components. For example, a cow producing 100 pounds of milk per day with a butterfat content of 3.5 percent will yield 3.5 pounds of cream after the milk is separated. A cow producing 60 pounds of 6.0 percent butterfat will yield 3.6 pounds of cream. Protein content can be similarly calculated. Feedstuffs may have a limited influence on some of these percentages, but the differences are mainly due to the genetics embedded within a given breed or animal.

Left: The Holstein is the most predominant milk breed. Most are black and white, but some possess a red hair color recessive gene.

Picking a Breed

Each breed of cattle has a long history of serving humans and was developed for specific purposes. It's up to you to pick one that suits your interests.

Ayrshire

The Scottish county of Ayr is the ancestral home of the Ayrshire, a breed that developed in the eighteenth century or earlier. Several strains of native cattle are believed to have been crossed with Teeswater stock, which was also used to develop the Shorthorn breed in England. The Scottish borders have a harsh climate, and the breeders developed the Ayrshire to thrive in such circumstances.

The cows became efficient grazers, able to survive on sparse land and less-than-ideal forages. Ayrshires are valued for their strongly attached, evenly balanced, symmetrically shaped, and quality udders. These are positive assets whether hand-milking or using a milking machine.

Ayrshires are typically red and white, but this can vary from white with red spots, red with white spots, all red, or all white bodies. In some cases, the red can be very light red, deep cherry red, mahogany, brown, or a combination.

Ayrshires can reach an average mature size of 1,200 to 1,300 pounds. They are strong, rugged cattle that adapt to varied management styles and are excellent grazers. Their calves are hardy and easy to raise because of their vitality. They are not the highest producing breed, on average, but can easily reach 16,000 pounds of milk per year with averages of 3.9 percent butterfat and 3.4 percent protein.

Ayshire
Ayrshires can be confused with red and white Holsteins because of the similarity in colors. They make excellent grazers and adapt well to many different terrains and climates.

Brown Swiss

The Brown Swiss is considered the oldest of the purebred dairy breeds. They originated in Switzerland's rough, mountainous regions and are known for the stamina and vigor provided by their large frames and heavy bone structure. This durability suited them well living under the rugged conditions of the Swiss Alps region. They are solid in color and can vary from fawn to very light brown to dark brown or almost black. The Brown Swiss's nose, tongue, and tail are black, as are the hooves.

Brown Swiss are among the most gentle and docile of the dairy breeds and are easily managed. They adapt well to different environments and management situations, especially grazing. They are among the largest dairy breed with cows typically averaging 1,300 to 1,600 pounds at maturity. The decades-old joke that Brown Swiss cows give chocolate milk is not true; however, a typical Brown Swiss cow does have good milk components and produces about 16,000 pounds per year with a 4 percent butterfat and 3.6 percent protein average. Their calves are hardy and grow rapidly, and the bull calves grow well as steers.

Brown Swiss

Brown Swiss produce rich white milk high in butterfat and protein components. They are a docile breed, making them popular with children.

Guernsey

The Isle of Guernsey, the Channel Islands, and France are the ancestral homes of today's Guernsey cow. The breed was introduced to the United States in about 1840. For the first half of the twentieth century, they vastly outnumbered other cow breeds in the country. The breed was heavily promoted as producing high-butterfat milk with a high concentration of beta carotene, which gave the milk a yellowish color. There still are many people who believe this milk has inherent healthy benefits besides the fat and protein components.

Guernsey
Guernseys have increased in popularity because they adapt well to grazing programs. Their milk is high in butterfat and protein. They are excellent at converting forage to milk and have docile dispositions.

Guernsey breed numbers suffered when the fat content in milk became less important to producers and buyers and breeds that produced larger volumes of milk became more popular. But their numbers are again increasing because of their good grazing abilities, gentle dispositions, ease of calving ease, and efficient milk production.

Guernseys are easily recognized by their light fawn color, bordering on a light orange, with white markings. They are intermediate-sized cattle with cows averaging 1,100 to 1,300 pounds at maturity. The average Guernsey cow can produce 15,000 pounds of milk per year with 4.5 percent butterfat and 3.6 percent protein. They have an advantage over other breeds by consuming 20 to 30 percent less feed per pound of milk produced than larger breeds, such as the Holstein or Brown Swiss.

Holstein

Holsteins are the black and white cows you spot in pastures while driving down a country road. They were originally known as Holstein-Friesian because of their origin in the province of Friesland in the Netherlands and in the Holstein region of northern Germany. In Holland, they were bred to make the best use of the grass pastures available. As the Dutch Friesians and Holstein cattle intermingled, their descendants developed into the distinctive black and white colors still prevalent today. Significant importations during the 1860s and 1870s brought a flood of Holstein-Friesians to the United States. Some Holsteins possess a recessive gene for red hair color, and red and white animals commonly live alongside their black and white herdmates.

Holstein
Holsteins count among the largest dairy breeds in body size. They can produce tremendous quantities of milk.

Holsteins are known for their ability to produce large quantities of milk that can, in many cases, approach the butterfat and protein content of other major dairy breeds. The average Holstein cow produces about 18,000 pounds of milk per year with 3.6 percent butterfat and 3.2 percent protein averages.

Holsteins are among the largest breed in physical size, and mature cows can weigh 1,500 to 1,600 pounds. Because of their large frames, they often require a larger housing area for comfort and to best suit their size. Calves are typically larger in size than those of other breeds as well. This may translate into providing more routine assistance at calving time.

Although Holsteins don't thrive on poor pastures like other breeds and are not as efficient at converting feed to milk as the Jersey or Dexter, they can consume large quantities of forages and can be good grazers if they are given adequate pasture.

Jersey

Many people admire the Jersey for its large, beautiful eyes and attractive face. The breed originated on the Island of Jersey, a small British island in the English Channel off the coast of France. It is one of the oldest dairy breeds with a known ancestry of nearly six hundred years. Importations of this breed were made to the United States in the 1850s, and they became popular because of the high butterfat content in the milk.

Jerseys are adaptable to a wide range of climatic and geographical conditions and are found all over the world, including Denmark, Australia, New Zealand, Canada, South America, South Africa, and Japan. They are smaller in size but

51

more heat-tolerant than many of the other dairy breeds. The average Jersey cow weighs 800 to 1,200 pounds at maturity.

Jersey calves are smaller when born, making birthing difficulties less prevalent compared to other breeds. Despite their smaller size, Jersey cows can produce more pounds of milk per pound of body weight than any other breed. Some Jerseys can produce thirteen times their body weight in milk per year.

Jerseys range from light gray to light or dark fawn to black in color. Their markings range from white spotting to solid colored. They may have black in the tail switch and muzzle with a light-colored encircling ring, making them appear as if they didn't get enough sleep.

Jersey milk averages the highest components of the dairy breeds with 5 percent butterfat and 3.8 percent protein on average. This makes them appealing if you want to make large quantities of cheese or butter. The average Jersey produces about 15,000 pounds of milk per year.

Jerseys typically have an excellent udder shape, which makes them easy to milk. The breed has made a tremendous resurgence in the total percentage of all dairy cows in the country, owing to the increased demand for high component milk for cheese- and butter-making. They are excellent grazers and can thrive on medium to poor pastures as their maintenance requirements are lower than breeds of larger size. Jerseys are typically docile cattle and can easily become family pets.

Jersey
Jerseys are a small-size dairy breed. They produce milk high in butterfat and protein, excellent for cheese- and butter-making.

Milking Shorthorn
Milking Shorthorns have increased in popularity because they are adaptable to grazing programs. They calve easily and produce sufficient amounts of milk for a family.

Milking Shorthorn

Shorthorn cattle originated in northeastern England in the Valley of the Tees River, in close proximity to Scotland. They were imported to the United States under the name Milk Breed Shorthorns in 1783 and were sometimes referred to as Durhams. They became favorites of many farmers for their quality meat and milk and as a source of power for fieldwork.

The Milking Shorthorn breed and the beefier Shorthorn breed trace to the same origin. During the 1700s, separate bloodlines were developed: one that was leaner with good milking qualities and one that was thicker, blockier, and meatier. The dairy cows can reach 1,100 to 1,300 pounds at maturity.

In the past two or three decades, the Milking Shorthorn breed has made much progress in transforming a beefy dairy animal into one designed primarily for milk production. Milking Shorthorn cows average about 16,000 pounds of milk per year with 3.7 percent butterfat and 3.4 percent protein averages.

Milking Shorthorns are red, white, red and white, or roan with color variations of each. They are one of the most versatile breeds and can adapt to a wide range of management systems, climates, and regions of the country. They are known for their regular calving intervals, living long productive lives, and having a desirable quality grading carcass when the end comes.

For family cows, the Milking Shorthorns are excellent grazers that can produce a lot of milk from grasses and rough pastures.

Red and White
Red and Whites can include any cow with a red hair coat color, but most often they are Holsteins. Red and White cattle can also result from crossbreeding.

Dexter
Dexters are the smallest dairy breed and adapt well to most farm situations. They are attractive for their high butterfat and protein milk. *American Dexter Cattle Association*

Red and Whites

Dairy cows with a red and white hair coat are often referred to as Red and Whites. They may appear similar in color and hues to the Ayrshire, but they have been mostly descendants of black and white Holsteins that carried the genetic recessive red hair color gene.

The Red and White Dairy Cattle Association (RWDCA) was developed in 1964 as a place for dairy farmers to identify the red and white calves they were getting from their black and white cows. At that time, red calves were banned from being registered in the Holstein breed's herdbook. The RWDCA opened its registry to include all red and white dairy animals regardless of ancestry, known or unknown.

The Red and White cow has production characteristics similar to those of the Holstein breed (see pages 50–51).

Dexter

Dexter cattle can be perfect for a family homestead. They require less feed than other dairy breeds and are gentle around children, easy to handle, and very hardy. They turn pasture grasses into rich milk and lean meat. They are one of the world's smallest true breeds of cattle, not a miniature developed from a larger breed.

The Dexter is the smallest dairy breed with the average weight of a cow being 600 to 750 pounds and standing just 36 to 42 inches at the shoulder. They are roughly one-half the size of an average Holstein cow. Calves weigh about 45 pounds at birth, and the delivering cow needs little, if any, assistance.

The breed originated in southern Ireland and was introduced into England in 1882, and then to the United States at the turn of the twentieth century when more than two hundred Dexters were imported. Most were brought to Kentucky, New York, Minnesota, and, later, Connecticut.

The breed is predominately black in color but can also be red or rust. This coloration may reflect their ancestry, which is thought to include the Kerry and perhaps the Devon, both red breeds.

Because of their smaller size, Dexters require less pasture and feed than other breeds, making them excellent for grazing programs in hot or cold climates with limited available acreage. They are raised from Alaska to Florida. They do well outdoors year round, typically requiring only a windbreak, shelter, and fresh water with their feed.

Pound for pound, the Dexter can produce more milk for its weight than any other breed. They produce an average of 8,000 pounds of milk per year. They have excellent components with an average of 4.0 to 5.0 percent butterfat and 3.5 percent protein. They are capable of yielding one quart of cream per gallon of milk, which is ideal for making butter or ice cream.

Dutch Belted

The Dutch Belted breed, also called the Lakenvelder, has a distinctive white strip around the middle of the body separating solid-black front and rear ends. The breed is listed as a critically rare breed of livestock in North America with about two hundred registered animals in the United States. They have made resurgence in recent years among family cow owners because of their docile temperament and suitability to grazing programs. They can produce large quantities of milk from only pastures, with no supplemental grains needed.

Dutch Belted were sought after by kings and noblemen because of their peculiar and striking markings. They still inspire curiosity today with their visually striking coat pattern.

The breed originated in the Tyrol areas of Switzerland and Austria, and they were highly prized for their milking and fattening abilities. They began to flourish in Holland where they were noticed by circus magnate P. T. Barnum. He is

Dutch Belted
The Dutch Belted breed has a distinctive white stripe between two black ends. They are gentle animals with quiet dispositions and work well as family cows. *Lisa Guell*

said to have imported them for show purposes and later placed them on his New York farm, which is regarded as the beginning of the breed in America.

The Dutch Belted is not a large breed, whether discussing its physical size or its population. The cows are moderate in size and weigh only 900 to 1,300 pounds. Their availability may be limited because of the low numbers that exist in this country. They are considered an intelligent breed with friendly, quiet dispositions, making them easy to handle.

Although Dutch Belted average only about 10,000 to 12,000 pounds of milk per year, their milk has very high butterfat and protein percentages, ranging from 3.5 to 5.5 percent butterfat. Their milk is excellent for drinking, and it yields better in cheese-making than an equal milk quantity from other breeds.

The lower production volumes are offset by several advantages Dutch Belted have over other breeds, including smaller birth weights. Calves typically weigh only about 70 pounds, which aids in easy calving for the mother. Although smaller at birth, the calves are thrifty and aggressive eaters.

If you choose to own a Dutch Belted family cow, it is imperative that you breed her to a purebred Dutch Belted bull to help increase the base and numbers of the breed rather than crossbreeding her to a bull of another breed. You can be part of developing the history of this breed that is making a comeback.

Crossbred
A crossbred cow has mixed ancestry and can possess the best qualities of the breeds involved. The cow shown here is an Ayrshire/Milking Shorthorn/Holstein cross.

Crossbreds

Crossbreds are not a separate breed designation but rather the result of mating two breeds to produce offspring that potentially carry the best traits of each. There are a couple of advantages that typically result from crossing different breeds rather than using only one breed.

The first advantage is heterosis, better known as hybrid vigor; this is the measurable gain in vigor and vitality resulting from combining the genes of two different breeds over straight-bred parents. Purebreds, in comparison, can be plagued with genetic weaknesses that accompany years of inbreeding.

Another advantage of crossbreeding is simply that you are blending the best qualities of each breed used. A Holstein-Jersey cross, for example, might inherit the high milk production of the Holstein and the high butterfat components of the Jersey. The opposite, however, also has the chance of occurring: getting the least desirable qualities from each parent.

Some international dairy breeds are being used in the United States for cross-breeding purposes, including Norwegian Red, Red Angler, and Aussie Red. It is not possible to import live animals from these breeds, but semen from bulls of each can be secured for a crossbreeding program.

Crossbred dairy cattle can be identified quite easily; many will exhibit multiple color markings that are a result of their mixed ancestry. Often, this is seen in the color patterns on the legs, head, or tail.

Milk production from crossbred cows varies widely, depending on the inheritance they were lucky enough to receive. Crossbred cattle are often hardy and stay healthy because of the hybrid vigor effect. They can be as good producers as purebred animals, so their ancestry should not discount them from your considerations for a family cow. Their conformation should be more important in your selection than their color pattern.

Normande

Originally bred as a dual purpose cow for meat and milk in Normandy, France, the Normande has received attention as a dairy breed in the United States in recent years. The breed possesses several characteristics that have increased their adaptation into crossbred breeding programs, such as being good grazers, having high feed conversion rates, having high fertility, and calving with ease.

They produce high-quality milk, often testing an average 4.4 percent butterfat and 3.6 percent protein from 14,000 to 15,000 pounds per year. Their milk contains high levels of casein beta and kappa, which improve the curdling quality for cheese production. It is common to get cheese yields of 15 to 20 percent higher from Normande milk than milk from other breeds.

An average Normande cow weighs 1,200 to 1,500 pounds and gives birth to vigorous calves weighing 70 to 95 pounds. They have good mothering abilities and good fertility.

The Normande is a red and white cow with occasional areas of brown hair. The brown hair may have the look of tiger stripes, or brindles, mixed with red spots. While one color often dominates, there appears to be some degree of balance between the three different hues, and the coat tends to darken with age. Calves do not display their brindles until a few weeks after birth.

Normande cattle cannot be imported as live animals, but they can be imported as embryos and then raised here. Most Normandes in this country, however, are the result of an upgrade breeding program, where a Normande sire is used in successive generations.

Normande
The Normande is a medium- to large-size dairy breed that originated in France. They work well in grazing programs and have good mothering instincts.
Marcus Hasheider

Milking Devon

In the United States, the Milking Devon has been bred from the Devon that originated in Great Britain. It was originally a dual purpose breed—and, to some extent, still is—because they yield an outstanding carcass when their milking days are finished.

The Milking Devon has a red hair color that can vary in shade from light to deep red or chestnut. They may have a white tail or udder. They are a versatile breed and can thrive in cold or hot climates.

Mature cows are of medium size and can weigh 1,000 to 1,200 pounds. They are slightly smaller than a Holstein and a bit larger than a Jersey. They have few calving difficulties and make good mothers.

Milking Devons are known more for quality of milk than quantity, which makes them ideal for cheese- and butter-making. The Milking Devon produces Jersey-quality milk without the addition of grain to her diet. Butterfat of 4 to 6 percent or higher is common.

The Milking Devon can adjust her milk volume to meet demand, which is an attractive feature if you want to use her as both a nurse cow and family milk cow. This natural flexibility will allow you to include her in a once-a-day milking program.

The breed is listed as critically endangered. There are now about six hundred living registered Milking Devon cows in the United States so their availability is more limited than other breeds. But if you can find one for purchase, they are an intelligent breed with a docile temperament and respond readily to calm, kind treatment—the qualities of an ideal family cow.

Milking Devon
Milking Devons are gentle, docile animals. Their breed numbers are low but if you can locate a cow, she will adapt well to a grazing program in any climate.
Marcus Hasheider

American Lineback

The American Lineback takes its name from its unusual color pattern: black on the sides of its body with a white line down the back and along the belly. Their color can also be roan or white with red or black speckles on their side; black and white; red and white; or roan.

There are other color variations, which are a carryover from their Witrick, Gloucester, and Welsh cattle ancestry in England. The Witrick pattern has speckled or dark sides; a black nose, eyes, and ears; and divides into three major types of color patterns: White Classic, Dark Sided, and Dark Speckle. The White Classic Witrick, which has proven to be the most dominant pattern when crossbreeding with other Linebacks or breeds, is mainly white except for some colored sprinkles across the body. It has a dark nose, dark ears, and a dark outline around the eyes. The Dark-Sided Witrick has a dark solid color pattern on both sides of its body with a white line down the back and tail, and from the brisket to the end of the belly. The Dark Speckle Witrick has more speckles and equal parts of dark and white coloring.

There is also the Gloucester Lineback in the American Lineback's heritage. These have a solid black head, sides, and legs with a white belly and white garter coloring around the tops of the legs, along with a distinctive white stripe across the back from head to tail.

American Lineback cows are medium size and can range from 1,000 to 1,500 pounds. They can produce between 12,000 to 15,000 pounds of milk each year, with an average 3.5 percent butterfat and 3.8 percent protein. A national association was formed in 1985, and there are several thousand Lineback dairy cows in the United States today.

American Lineback
The unusual but attractive color patterns of the American Lineback appear in several variations. Shown here is the Dark-Sided Witrick. The solid dark color pattern is prevalent on both sides with a white stripe from head to tail, along with mostly white legs. Lineback cattle can be black and white, red and white, or roan, and with light speckling of either black or red. *Philip Hasheider, Joel and Lisa Guell Farm*

Making Your Decision

Personal preference, color, size, and availability may play a large part in your decision of which breed of cow you choose. While these are important considerations, make sure the cow you choose is a physically sound, healthy animal. You will not be happy with an unsound cow simply because you liked her color or she happened to be the only one available. Take the time to study which breed is attractive to you and then go out and look at some. Then you will be prepared to make a good decision that you can live with.

The cow you bring to your farm will become a part of your daily routine. Take the time to choose the breed and cow that is right for you and your family.

CHAPTER 4

Housing and Fencing
Home Sweet Home

■ ■ ■

IT DOESN'T TAKE an expensive barn to shelter your cow. You may already have a small shed or garage that can easily be converted to suit her needs. Since you only have one cow, the amount of space needed is different from having two or three.

Your cow will need to get out of cold rains, bitter winds, or blistering heat. Of the two extremes of hot or cold, it is typically harder for a dairy cow to handle heat and humidity than cold conditions, provided she can find relief from the wind.

Heat

The reason a cow experiences difficulty with heat and humidity is because of the way her body works. A cow's body is like a fermentation vat that creates a large amount of heat while the digestive processes are working. In times of high heat or humidity, she has a hard time expelling the heat fast enough to keep her systems cool, especially if she can't find shade.

If you live in an open area with few trees, build a simple structure to provide shade, such as canvas or metal sheeting nailed to the top of posts set in parallel rows. Anything that blocks the sunlight during the hottest part of the day will help.

Portable barn fans are available in different sizes. You can move them between areas where your cow is housed. This flexibility makes them as attractive as stationary mounted fans.

Left: Gates allow a cow ease of entrance and exits from fields surrounded by fencing. Gates can be made of wood, metal tubing, or wire welded to an iron frame and anchored to solid posts.

Trees serve as an effective sun screen for your cow. However, they may act as lightning rods during thunderstorms. You need to be careful if using them for shade during stormy times of the year.

Humidity is one of the most difficult climatic conditions to control. Setting up an air conditioning system in a small area of your stable or garage may be feasible if you think it will be needed for only a few days. Large fans can be useful to create air movement, offering some relief, but typically it is still warm air being circulated.

Cold

Cold temperatures typically don't have as great of an effect on a cow's health as heat does. She can generally create enough body heat to stay warm in bitterly cold weather. There are a few factors, however, that impact her ability to manage the cold.

Bitter winter winds make it difficult for your cow to maintain adequate body temperatures if she is continually exposed to them. The greatest hazard for her in that type of weather, however, is the potential for frozen extremities, such as her four teats. Although they are resilient structures with blood flowing through

A manure pack can be used during the winter. It adds comfort as a soft bed. To be most effective, bedding must be added every day to keep your cow dry. Be sure there is adequate ventilation.

A metal windbreak can be used if no other shelter is available. Metal sheeting or corrugated steel panels are solid enough to withstand most high winds. *Marcus Hasheider*

Trees can serve as an effective windbreak for your cow in cold winter conditions, especially if they are close to your buildings.

them for warmth, they can be likened to your fingers where the tips become frozen first and slowly this depresses adequate blood flow to keep the tissues warm. Severe frostbite can occur in cows just like in human extremities. A simple windbreak made of wood, trees, metal, or even the leeward side of a building offers her protection.

Another threat in cold weather is water in the form of ice or snow. Just like you, a cow can slip on the ice and injure herself. Even more threatening is the moisture she brings into the stable area. Lying on wet, moist bedding in cold weather is a recipe for respiratory illnesses. The wetness has a chilling effect on her body. Some dairy producers use bedding packs during the winter. This involves adding dry bedding on top of the existing bedding that may have gotten wet. Rather than cleaning out the manure pack every day, they add to it, and as the chemical reactions between the manure components and moisture occur, it generates low-grade heat, providing a comfortable bed for their cows. It is similar to a compost pile but is maintained in the stable until warmer weather, when it is cleaned out.

Your cow will eat more in the winter to compensate for the energy she burns up maintaining her core body temperature, but she has little defense against the wet cold. Ice and snow can impact her production levels if she does not get enough to eat; however, neither will seriously threaten her health if you maintain a dry stable.

Generally, the conditions under which you would prefer to comfortably milk your cow are also the conditions she would like to live in. Use yourself as a gauge, and you can better sense what will make your cow comfortable. All weather patterns pass, but you need to take extra care of her during the extremes so that she survives them as well as you do. She doesn't need to live in your house, but you can provide her with a comfortable shelter and bed.

Planning Simple Shelter Structures

You may be able to pasture your cow the year round in mild climates. In that case, her housing needs are minimal: shade and wind protection. In more extreme climates, a structure will be needed to provide some protection against wind, rain, and sunlight. Her most basic space needs for a stall depend on her size. A 1,000- to 1,400-pound cow will have adequate space in a 10x12-foot box stall. Smaller cows will have plenty of room in this size pen and can even do with less if that is all that is available.

Besides comfort issues, there are three main factors that should be considered in how you house or stable your cow. These will be approached in a "backward" manner.

Manure Removal

Consider manure removal from a bedding area first. Although this is discussed in greater detail later in this book (see chapter 6), you will need to remove about one hundred pounds of manure and urine every day if she resides completely inside and you clean her pen daily. Stabling her in an area or in a manner that makes manure removal easy will lessen your workload. Manure removal is not a great problem if your cow has access to pasture for most of the day and only returns to the pen for milking. She will spread the manure herself as she grazes across the field. When not in the pasture, she will deposit her manure and urine in the only place accessible, her pen.

You may be able to adapt the inside of the barn or shed where you keep your cow to help make manure removal easier. If it is a small barn with stanchions or

A structure for summer shade can be built, or an old shed standing in a pasture can be used if it's available. Any structure should be positioned to provide the maximum amount of shade in relation to the late morning or afternoon sun.

a tie-stall with a gutter behind, this is an easy way to handle the manure. A gutter is a deep trough behind the cow stall that may or may not have a chain with attached paddles that move the manure along the cement surface, powered by an electric motor system. By flipping a switch, you are able to clean the gutter as the chain moves the manure and deposits it outside the barn, whether into a manure pit, manure spreader, or on a stack for later hauling.

If there is no chain, you can shovel the manure from the gutter into a wheelbarrow and haul it outside. Either way, a gutter provides space to store the manure for a short time without having to handle it every day.

Another option is to build a pen or stable to fit your specifications and needs that makes it easy to remove the manure. If you build or remodel a pen, whether it is in a shed, a lean-to, or part of a garage, consider allowing for a large gate that swings open so you can get in with a skid-steer loader or some other mechanical lift to clean the pen. This large gate will make cleaning easier. Packed manure is often difficult to remove from a pen with a pitchfork. Having a narrow entrance gate

This small milking parlor has a gutter to catch manure. It can be manually cleaned with a pitchfork and shovel. If the gutter has an operable cleaner chain, it will make the job easier.

for your cow will restrict your movements when you need to clean out her pen. In that case, you will be removing most of the manure by a pitchfork and hauling it to a pile or pitching it into a manure spreader.

During winter, it is sometimes more convenient to keep adding dry bedding and create a bedding pack for your cow to lie on rather than to clean it daily. Having a way to get at that pack will minimize a lot of labor later. Whatever you choose, just remember that the design of your cow stable should take into account how you will remove the manure.

Pens or Stalls

One advantage of a box stall or open housing over a stanchion or tie-stall is the freedom of movement your cow will have outside of milking time. In a

A wheelbarrow and pitchfork are basic manure handling equipment. Larger equipment may be necessary for hauling to fields, but around the farmstead these will accomplish most jobs.

tie-stall, she is tethered by a neck strap and chain. In a stanchion, she is restrained by metal bars on each side of her neck, which allow movement but are still too narrow for her to pull her head out until you release the lock at the top; then she can pull her head back through and exit the stall.

Tie stalls and stanchions are humane forms of restraint, but they do not allow the freedom of movement that a box stall does. With one cow, you can easily use gates to create a pen or a large stall even if one doesn't exist in the barn or shed.

Unless she is a very large cow in excess of 1,800 pounds, your cow will have enough room in a pen measuring 10x12 feet. Smaller cows can do with less and still be comfortable in a pen measuring 10x10 feet. The smaller the pen, however, the quicker the manure piles up around her and needs to be removed or bedding added.

Arrangement of the stable should include where you plan to feed your cow. This is not a problem in a stanchion or tie-stall because she is fed in front of where she is tied. If using a stanchion only during milking time, her feeding can take place while you milk or after you're done and she goes back to where she stays the rest of the time.

Whether keeping her in a box stall or stanchion, feeding your cow away from the area where she lies down will keep the bedding cleaner and dryer and will reduce the amount of manure in that area; she's moving it for you.

You can more easily clean a pen if it has a large entrance gate. It is easier and faster to use mechanical means to remove the manure and bedding.

It doesn't take a magical design to construct the feeding area away from where she lies down, just do some thinking before you build it. Besides, the movement and exercise to reach her feeding area is good for her legs and feet, especially if she is confined during the winter.

Access to Water

Plenty of drinking water is essential for your cow in both extreme and normal weather conditions. A cow typically puts 85 percent of her water intake into her milk while the rest is used for body maintenance and function. A dairy cow can drink as much as 15 to 20 gallons of water a day in hot weather, or 120 to 160 pounds. That's a lot to carry in pails so many people hook up a water tank with a float valve, allowing a cow to drink what she wants. As she does, the valve opens automatically to replace the amount taken out. This laborsaving device can be used with many different size tanks.

Your cow requires about as much water during extreme cold periods as she does during hot temperatures. In the winter, she can't break through ice so you will need to provide water by supplying it to her several times a day or using a water tank that's available whenever she wants to drink. It is a myth that cattle can survive only on snow for their water supply during the winter. They use more energy and liquids to transform the snow into water than what they gain from it. The result is that cattle dependent on snow for their water supply become dehydrated and die quicker from a lack of water during a snowstorm than from a lack of feed.

A stanchion is a humane form of cattle restraint. The two cows pictured here are hand-milked daily. The setup is typical of a small family farm.

A tie-stall is a similar form of restraint to a stanchion except the head lock is removed. It is replaced with a neck strap to which a stationary chain with a snap ring is attached.

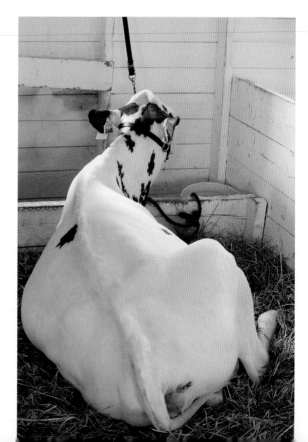

Shorting a cow's water supply will lower her production, especially in hot weather. Fresh, clean water should be constantly available by a drinking cup, water tank, tub with a float valve, or any other means available. Otherwise you will need to offer her water eight to ten times a day—prohibitive for most people.

Freezing winter conditions pose several challenges to providing a steady supply of water. Using electric heat tapes that wrap around water pipes will help keep lines open. Heating elements can be used in metal water tanks to keep areas open where she can drink. Water and electricity can make a deadly combination; be certain these elements meet all electrical standards and that the electrical system is adequately grounded.

A water tub can be fitted with a float valve or you can fill it with a hose. The farther away the tub is from the water source, the less water pressure there is to transport it from your well.

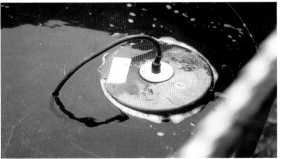

In cold climates, water tanks can be fitted with heating elements. This will keep the water open during freezing weather. These elements must be properly grounded and the electric cords positioned so your cow cannot chew on them.

Good Ventilation

Your cow needs fresh air but not drafts. She processes about 75 pounds of feed each day and 100 pounds of water to produce her milk and maintain body functions. The moisture she expires from these processes needs to be removed so that she does not develop respiratory problems. With just one cow, this is unlikely to be a problem. If other animals are also housed with her, the windows should be located on the wall so that the air movement is above them and not at their level. Sheds, lean-tos, or other structures that have an open side eliminate any need for ventilation equipment; nature will take care of it for you.

If you are building a new structure or devoting a corner of a barn to her, choose a side away from the prevailing wind for ventilation windows. This consideration is especially important if she is to be on a manure pack during winter. The heat from the pack also creates ammonia vapors that can affect her, and proper air movement will avoid complications.

Hutches or huts are separate compartments to house calves. They should be placed so the open end faces south in winter. Move them to a shaded spot in summer. *Nigel Cattlin, Visuals Unlimited, Inc.*

Housing for her calf is similar in the basics for a cow. The space should be dry and draft-free, with proper ventilation. Depending on your milking stall plan, you may even house the calf in the same pen as her mother. If you wish to keep the calf away from its mother, you can use a calf hut, which is like a single room apartment for calves and allows you to easily care for it. The huts resemble igloos and have a similar effect. In extremely cold weather, a calf will lie in the back, away from the opening, and snuggle into the dry bedding.

Small pens built inside a barn or shed can house calves. Feeding is done in the front with pails. Bedding must be frequently added to keep the calf dry.

Permanent and Temporary Fencing

Good fences make for good neighbors. There's more than poetic truth to that statement. Fences have marked the boundaries of farms for generations, giving the owners containment for their livestock and protection from animals on adjoining farms. Having good fences is important, not only as a mark of property lines, but also as a way to humanely enclose your cow. A fence will keep her in your pastures and your neighbors' animals out.

Whether you raise only one cow or several other species with her, your farm will likely require two different kinds of fences: permanent and temporary. Permanent fences are intended to last for many years with minimal repairs and should be constructed with sturdy materials. These are typically perimeter fences or fences along streams or waterways. Temporary fences are intended to last only a short time and are often moveable. This mobility allows them to be constructed from less expensive materials that are more flexible than permanent fencing.

A perimeter fence is more important than an inside temporary fence as it is your last line of defense if your cow happens to escape her shed or pen.

A shed can easily be converted to a stall or pen barn. Keep the design simple and it will work well for a single cow.

Step-in posts are often used for temporary fencing. These are easy to insert and remove. They have multiple levels of clip holders to attach the fence wire depending on the height required.

If it is not possible to construct a permanent fence around the entire perimeter of your farm, consider building those sections that will be most useful to your farm plan and your cow.

You may choose to tether your cow to graze specific areas she might neglect, such as roadsides or corners of fields. The method of restraint is easy: Simply drive a stake into the ground with one end of a chain or rope attached to it and the other end attached to the cows' neck collar. The chain may be 20 feet long, which lets her eat in a 40-foot radius. The stake can then be pulled up and driven in at a fresh spot the next day. While this system can avoid wasting pasture ground, it also carries some risk of the cow getting tangled up in the chain with no one around to help her. Also, staking a cow in an area with no shade or limited access to water during the day will defeat the purpose and endanger your cow.

Permanent fence materials most appropriate for grazing cows include barbed wire, woven wire, and electric wire. Other materials, such as high-tensile wire and cable wire, work well for confinement areas, such as corrals, around barnyards, and in loafing areas.

Types of Wire Fencing

A standard **barbed wire fence** uses three to five strands of wire with posts spaced 10 to 12 feet apart. A five-strand fence typically has spacing of about 10 inches between each strand and will be sufficient to deter cows, calves, sheep, and goats from trying to push their heads through. The fewer wire strands in the fence, the wider the spacing between them. This makes it easier for smaller animals to squeeze through or for large animals to push their heads between the wires to get grass on the other side of the fence. A three-wire fence may be adequate for a single family cow and would cost less to construct.

Woven-wire fences consist of a number of horizontal lines of smooth wire held apart by vertical wires called stays. The height of most woven-wire fencing materials ranges from 26 to 48 inches, which on its own is too short for cattle, but it should be sufficient when you run a single strand of barbed wire about 4 inches above the topmost woven-wire strand.

Woven wire is often more expensive than other types of fencing materials because of the additional metal used in its construction. It can cost two to three times that of barbed wire for a similar distance while having approximately the same lifespan. However, if you have multiple species on your farm, such as sheep, goats, or pigs, then woven wire may be worth the cost as an added deterrent. You may also cut costs by only using it in a specific area designated for permanent pasture and using cheaper materials for temporary enclosures.

A four-strand barbed wire fence is a strong structure that works well for many species. These are most often built along farm boundary lines or permanent pasture areas.

Above: A woven wire fence consists of several horizontal lines of smooth wire held apart by vertical wires called stays. These are strong, durable fences. A row or two of barbed wire may be attached to the posts above the woven wire to aid in deterrence. *Marcus Hasheider*

Right: Electric fencing is used in areas separated from large portions of the field or pasture. The wire is attached to plastic insulators. This keeps them separated and ensures an unbroken electrical circuit.

Electric-wire fences are more flexible in their uses and materials and can be used both for permanent and temporary fencing. As a moveable fence, it can be made with one or two strands of smooth wire or a poly tape that has small wires woven into it to carry the electric charge. The poly tape is more flexible and easier to handle, and it moves from one location to another better than smooth wire. In either case, the wire is energized by an electric controller that receives its source from a standard farm electrical outlet or a solar-powered pack. This source sends a low-energy pulse through the wire at short intervals. Both need to be grounded to complete the circuit. An electric fence has several advantages, including being easy to move from one area to another so that you can change access to different parts of the pasture when needed.

There are a few drawbacks to an electric fence. If your cow is unfamiliar with this type of fencing, she may need training. But after receiving a quick snap from touching the electric pulse-driven wire, she is likely to stay clear of it. And of course electric fences don't deter escape if the current no longer runs the length of the wire or when vegetation grounds the wire. You will need to maintain the fence charger and keep the wire clear of any plants to be effective.

There are many booklets available that will explain how to build proper perimeter or temporary fences, and you should consult them for further information. You can also learn about the variety of fencing materials and plans available by contacting your county agriculture extension office or a fence manufacturing company.

CHAPTER 5

Feeding
What's for Dinner?

■　　■　　■

A FULL PLATE makes for a happy cow. Cows eat a lot. Keeping that plate full can keep you busy. Providing pasture for most of the growing season is the easiest way to feed your cow. It's also the cheapest because grass is pretty much free food. It doesn't cost a lot to grow it. Pastureland should be sufficient to meet her needs for most of the year if you have adequate rain and plant growth. You can either purchase what she eats during the winter months or you can harvest your own hay crop and store it for winter feeding if you have the space.

Your cow has a daily food requirement like you do, only more of it. Her diet has to support multiple demands: producing milk, body maintenance, and supporting a developing calf if she is pregnant. During her dry period, when she is not milking, her nutrient intake goes into the calf, into her body maintenance, and into storing reserves in anticipation of coming back into milk production. The amount of milk she will give is greatly affected by what she is fed or eats on her own, especially if she is a breed with high milk production potential. Whether you choose to feed for maximum production or not will depend on your personal goals. If you want to process multiple dairy products for home use, you may want higher production; if you want just a supply of raw milk, then you may want lower production.

Your cow is a ruminant, meaning her stomach has four chambers. The chambers are fermentation vats that break down plants and grains into carbohydrates, proteins, amino acids, enzymes, and other components needed for growth, maintenance, and milk production.

Cattle are amazing creatures because this specialization gives them the ability to absorb nutrients and break down grasses through a process called acidification. This process lets your cow extract nutrients from low-quality feeds and effectively utilize plants that other animals can't. They become a conduit for plants and grasses not edible by humans and convert them into meat or milk that humans can use.

Left: The upper loft, or haymow, of a barn is often used to store hay. When needed, the bales can be tossed outside or through a chute into the barn. Loose hay can be pitched out into a wagon or cart.

The key to the ruminant digestive system is a symbiotic relationship between the microorganisms that exist in the fore part of the stomach. These bacteria and protozoa have the capability to convert solid plant material into enzymes a cow can use. As these microorganisms work together, they produce enzymes called cellulase, which cannot be produced on its own by a ruminant. Cellulase enzymes break down the cell walls of plant materials, which releases the plant's fatty acids into the digestive system. These fatty acids are then absorbed by the cow and make a significant contribution to her overall energy needs. As the other plant materials pass into the intestine, amino acids, lipids, carbohydrates, and other enzymes are absorbed in much the same way as in other animals. The presence of microorganisms to aid in two different digestive systems allows the cow to exploit cellulose as an energy source that other plant-eating animals cannot. This makes cattle ideally suited to eat crops produced on poor soils that may be indigestible or unsuitable for other species.

Your Cow's Digestive System

Your cow's stomach has four chambers: the rumen, the reticulum, the omasum, and the abomasum. They work together as a specialized system. The rumen is the largest chamber and the first place where the grass eaten begins its trip through the digestive system. The rumen is like a slow-moving tumbler that gets filled with chewed and half-chewed materials where they get mixed with saliva as the cow regurgitates, chews, and swallows these materials several times.

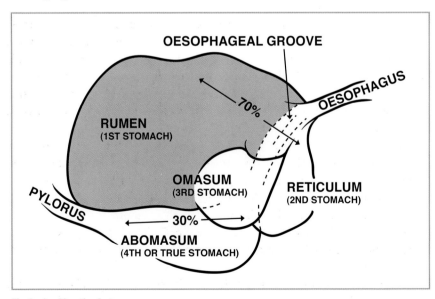

The Bovine Digestive System

When a cow lies down and starts "chewing her cud," she is using the chewing process to grind the food that has been in the rumen into smaller portions. This material is mixed with her saliva as it is crushed and chewed. This use of her saliva is one reason a cow drinks large amounts of water during the day. The crushing and grinding of the cud allows more surface area of the material to become exposed to the bacteria in the rumen. While it may seem that your cow is not doing anything but chewing grasses swallowed hours ago, she is initiating the fermentation process.

As the fiber breaks down into cellulose, the bacteria and protozoa then convert the cellulose into glucose, which is used, in part, by the microorganisms to feed themselves. While some microorganisms escape the rumen and pass through the other chambers, most stay behind to work on the newly ingested plant materials.

The fiber that is broken down immerses itself into the rumen's liquid content—the saliva and stomach acid—and then passes through to the next chamber, the reticulum, and then to the omasum, where the water is removed. All through this process, the materials are constantly being churned to mix the liquids, solids, and bacteria to keep the fermentation process brewing.

As the water is being removed, the mixture moves along to the final chamber, the abomasum, where it becomes digested much like it would in the human stomach. The abomasum—unlike the rumen, reticulum, and omasum—does not absorb nutrients. It prepares food for enzymatic breakdown and absorption in the small intestine.

The absorption of nutrients continues in the intestine until the unusable portion is finally expelled from the rear end of your cow as feces, or manure. This digestive process produces acetic acids, propionic acids, and butyric acids, which are volatile fatty acids that give ruminants their energy. The volatile organic compounds that are expelled can be broken down through composting and deposited into your garden to complete the cycle that started with your cow eating a plant.

This information may be more than you expected to know about how your cow turns grass into milk. However, it shows how intricate this system is to her well-being and production; anything that interferes with this process will ultimately affect both. Not eating for one or several days will cause these bacteria to decrease in numbers, which in turn affect their output and support of her body functions. A sick cow will confront more challenges than just recovering from what ails her; she needs to regenerate her stomach bacteria to recover her body processes.

Happy cows are cows chewing their cud. It's a beautiful sight of contentment and good health.

Grass for Feed Is What She Needs

Most of the nutrient and energy needs your cow requires for growth and production can be obtained from good-quality grasses alone. Grasses, clovers, and alfalfa are the most abundant natural sources available for feeding cattle and are the least expensive crops to produce and harvest.

In many areas of the United States, grass grows plentifully and can be used as the sole or major source of feed. There are many varieties of grasses and legumes, such as **clover** and **alfalfa**. You can seed for pastures if none exist on your acreage, or you can inter-seed and increase the production of your existing pastures.

It's a good idea to have different grasses growing in your pastures. This allows for more flexibility as climate conditions change throughout the year. Some grasses grow best in cool weather, others in warm weather. There are grasses that thrive on the mid-summer heat, which normally slows down growth of many others.

You can reseed or inter-seed grasses to increase the diversity and length of the growing season. **Timothy**, **clover**, and **bluegrass** prefer cooler temperatures and grow best in the spring and fall. Legumes such as **alfalfa** start growing later in the spring when it's warmer, but have a strong, uniform growing pattern during the summer season. Few grasses or legumes grow well in dry conditions, except for **bluestem** and **switch grass**; during dry conditions, most other grasses slow down or stop growing but recover when sufficient moisture returns.

Your cow doesn't need to eat a wide range of grasses to be healthy. It's just good management to incorporate different species in your pastures.

How Will You Feed Your Cow?

Decide how you will feed your cow. Then you can best calculate her nutritional requirements. Here are some questions to ask yourself first: How much time do I want to commit to producing feed for her? Can I buy her feed more cheaply than harvesting it? Can I hire the harvesting done versus doing it myself?

Basically, it's a balance between what you feel your time is worth and what the benefits are of hiring others to harvest for you.

If you don't have the acreage for your cow to graze as well as have enough grasses left to harvest, then you'll need to purchase feed to get you through the winter months. Good quality feed usually can be purchased locally, but you will need to plan ahead. Make arrangements with a neighbor before the growing season to reserve a source of feed for your cow. Another option is to purchase through a feed wholesaler.

You can make your own hay using a tractor or haybine, or you may decide to hire the hay cut by a neighbor and do the baling yourself. Large haybines are expensive, and hiring the work done may be more economical than owning the equipment.

Purchasing equipment to harvest feed typically involves a fairly high financial investment. You need the minimum of a tractor, hay cutter, rake, baler, and wagon to make hay. It would take several years to recover the cost of this type of investment when compared to hiring the work done or purchasing feed. It's difficult to justify the expense of all this equipment just for one cow. Even if you're mechanically inclined, buying old, inexpensive equipment can be like buying a rusty used car. It will likely take a lot of effort to get it working properly. And sometimes the total cost of the repairs exceeds the original price.

What's your return for the cost invested in feed? Calculating a feed cost–to–milk ratio helps you determine this. If you feed your cow for maximum production and then don't make use of the milk, it's a lost income opportunity. To avoid this, the milk can be processed by you into dairy products with a longer shelf life. Or, you can raise other livestock that make good use of the milk, such as pigs. The high fat and protein levels in milk provide pigs with a terrific diet to grow quickly.

A rake turns the flat row of hay into a windrow. It rolls the hay into a long single thread that can be easily picked up by a baler.

Dry hay can be handled in several ways. It can be stored as loose hay, in small square bales, or in large square bales, or in large round bales.

A hay baler picks up the windrow and directs it into the baling chamber. The plunger is driven through the chamber and compresses the hay into the rectangular shape of the bale chute. Bales are separated by twine string at set intervals that is automatically tied by a knotter.

Bales are pushed out behind the baler as it passes along the rows of raked hay. You can use a wagon to go around the field and pick them up later. Or if you have the help, you can have someone load the hay on the wagon as you are baling.

A round baler makes large bales that need to be moved with proper equipment, such as a tractor with a front- or rear-end loader or a skid-steer loader.

Alternatively, you can decrease your cow's milk production by lowering her feed levels. You can do this without sacrificing her health. This is a great option for a family cow, just keep in mind a few caveats.

One caveat is that it's easier to achieve and maintain high production at the beginning of your cow's lactation rather than later on. If you drop her production early, it's difficult to increase it later if circumstances change and you want more milk. By that point, she will have accommodated herself to the diet that supports her lower production, at least until her next lactation starts. You might get a small increase in milk by supplementing with grains, but she will typically not reach her former level of production at that point. Make sure you really want to reduce your cow's production before doing so.

Grains, such as oats, wheat, and soybeans, provide a high protein boost. However, grain is a concentrated source of protein, which means that feeding your cow more than she can utilize is wasted money. Ground shelled corn is often supplied as a carbohydrate, which is used in the fermentation process and broken down into sugars as an energy source. Failing to meet the energy needs at peak production, especially in early lactation, causes as much depression on those levels as a protein shortage in her diet.

Minerals can be offered free choice for your cow to access them anytime she wants, or they can be mixed into any grain being fed. Minerals help bring the calcium-phosphorus levels into balance and add trace minerals that may be lacking in the soil and therefore her pasture feed. A soil test of your pastures will identify any nutrient deficiencies. These can be done by a local agronomy cooperative or fertilizer business. You can also run a feed test of your hay after it is harvested or purchased to determine the nutrient content and then supplement what's missing. This is referred to as mineral balancing and can be done with the help of a local feed store or feed company representative. Feed companies may also offer soil testing as part of their business services.

Vitamins are usually contained in anything your cow eats. All leafy forages contain vitamins A and D. Sunlight directly provides vitamin D. Unless your cow is not exposed to sunlight in winter at all, she probably won't need it added to her diet. A standard mineral supplement can supply any she may otherwise be lacking. A mineral salt block is a nutritional necessity because most grass-fed diets are deficient in certain essential nutrients, including salt.

How Much Does She Eat?

Your cow may eat a lot. How much will she eat in a single year? Large breeds consume more feed than smaller ones. Some breeds, such as the Jersey, typically produce a richer fat and protein milk from the same amount of feed as a Holstein but with less volume. The total amount of feed you need to plan for is determined in part by the cow breed you choose.

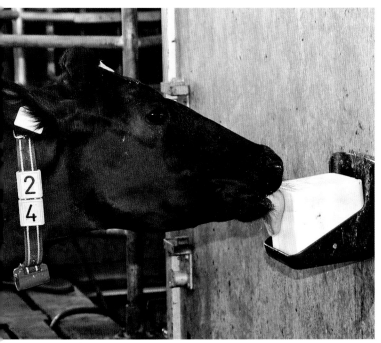

Trace minerals are an important part of your cow's diet. These are easily added by using a trace mineral salt block.
Shutterstock

An average dairy cow will consume between 2 to 2.5 percent of her body weight in dry matter each day (or 4 percent in total intake). The percentages hold true whether she is a large cow or a small cow. This is usually expressed in dry matter (DM), which means that the water has been removed to make calculations consistent, and it is the way in which most rations are figured. This figure can be reached two ways: by her weight and pasture feed, or pounds of milk produced yearly. Here are the equations needed to calculate how much dry matter she requires daily, and how this translates into actual feed consumed.

Example 1

A cow weighs 1,000 pounds and is on pasture.

Body weight × 2.5% = Dry matter required per day
1,000 lbs. × .025 = 25 lbs. DM

Grass is about 85 percent water, or 15 percent dry matter.

Dry matter required per day ÷ 15% = Grass required per day
25 lbs. DM ÷ .15 = 167 lbs. grass

Your cow will need to consume about 167 pounds of grass per day to meet her nutrition and maintenance needs.

Under normal circumstances, a healthy cow will eat until she is satisfied. You can weigh her feed with a scale to track how much she eats each day beyond what she gets in her pasture grazing. The scale can also be used to weigh grain or her milk.

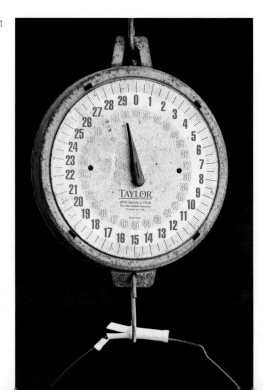

Example 2

A cow that produces 16,000 pounds of milk in one year (52.5 pounds per day for 305 days) will consume 4.5 tons of forage dry matter per year. There are 2,000 pounds in a ton.

Dry matter in tons × 2,000 pounds = Dry matter in pounds ÷ Days in a year = Dry matter pounds required per day

4.5 T DM × 2,000 lbs. = 9,000 lbs. DM ÷ 365 = 24.6 lbs. DM

Due to moisture content, 1 pound dry matter is equal to 3 pounds corn silage, 2 pounds of haylage, or 1.1 pounds dry hay.

If using only dry hay for feed, multiply the dry matter figure by 1.1 to find out the pounds of dry matter required per day.

Dry matter required per day x 1.1 = Hay required per day

24.6 lbs. DM x 1.1 = 27.1 lbs. hay

This formula can be used whether you want to milk a cow producing over 50 pounds per day or not. Any cow producing less than that will require less pounds of forage. Although these calculations are at the upper levels required, they give you an idea of the amounts you will need to sustain your cow through the year. Variables that may affect these average numbers include body size and climate conditions.

How Much Pasture Do You Need?

Using a pasture to feed your cow requires some knowledge of growing grasses so there is a sufficient amount for her to eat. Well-managed pastures provide more than enough plant materials for her to eat and for you to harvest for later use, if you have several acres available.

You've calculated how much she will eat during a year. You can also calculate how much of that can be produced from your land. If you only have your cow's pasture available for both grazing and making dry hay, it will be more of a challenge to acquire all you need for the year. If you have other acreage available other than her pasture, this will not be a problem.

Your cow will be eating over 160 pounds of plant material from your pasture each day. She does this in numerous eating episodes throughout the day, not all at once. During the flush growth of early spring, it is unlikely that she will be able to eat all the pasture areas completely. That would be a good time to harvest the excess as dry hay before it becomes too mature and its quality drops.

After harvesting the pasture for winter feed, you can put her back out; if weather conditions are favorable, you should see a quick pasture regrowth to supply her daily needs.

There are several handheld wands that can measure the pasture plant density with great accuracy to tell you how many tons per acre are available for grazing or cutting. Your county agriculture extension agent or a local feed company representative may have one to help you accurately determine your pasture growth.

A good pasture will typically produce from 3 to 5 tons of forage per acre per year.

Example 1

Say your cow is on pasture for six months of the year:

Days on pasture x Grass eaten per day = Grass eaten per year (in pounds)
180 days x 160 lbs. = 28,800 lbs. of grass

Grass eaten per year (in pounds) ÷ 2,000 pounds per ton = Grass eaten per year (in tons)
28,800 lbs. ÷ 2,000 lbs. = 14.4 T of grass

Grass eaten per year ÷ 4 tons per acre (average) = Acres needed for the year
14.4 T ÷ 4 T = 3.6 acres

This is the size acreage your cow will need for pasture for the six months she is pastured. You will need to feed her hay or other feed during the non-pasture winter months.

Example 2

If you've already established that you cow needs 25 pounds of dry matter per day and you want to feed her straight dry hay year round:

(Dry matter in pounds per day x 1.1) x 365 days = Amount of hay required per year
(25 lbs. DM x 1.1) x 365 days = 10,038 lbs. of hay or 5 T of hay per year

Dry hay is a good option if you don't have enough acreage to support your cow for the full year on pasture. Other feeds that may be available include haylage, which is fermented plant materials and can be made from alfalfa, oats, wheat, or barley; corn silage (although this is not high in protein); or vegetable refuse such as beet pulp, brewers grains, or canning peas vines. Dry hay has the advantage of being easily moved from one location to another. It can be purchased in different weights depending on your needs. It can be purchased in round bales weighing between 800 to 1,200 pounds. Or it can also be purchased in small bales between 50 to 75 pounds, or in large square bales of about 1,500 pounds each. The small bales can be handled manually, but the larger round and square bales will need a loader to move them unless they are opened and the hay is moved in small amounts.

Left: Large square bales weigh approximately 1,500 pounds each. They are easily stacked together because of their shape. It takes a large baler to make this size bale.

Below: Round hay bales are an alternative to square bales. They weigh between 800 and 1,200 pounds.

If you plan to feed straight dry hay and you've calculated that it will take approximately 10,000 pounds, or 5 tons, per year to feed your cow, then it will take ten round bales of 1,000 pounds each or six to seven of the 1,500-pound large square bales to feed your cow through the year. If you pasture your cow for part of the year, you can deduct the amount fed through the pasture season. You can also decrease this amount if you supply a grain concentrate as part of her diet.

Buying Hay

If you purchase dry hay bales, take a look at the inside of one. Pick a random bale in the lot and ask that it be opened. Check inside for any abnormalities in quality, such as mold. You can ask for a hay sample analysis. (Hay prices are based on quality, protein content, total digestible nutrients, and availability.) This report should tell you the quality of the hay, but it won't tell you if it contains mold.

Do not buy any hay that is moldy or excessively dusty. Mold results from improper or poor harvesting conditions where the moisture level in the plants is

Do not feed your cow any moldy hay, silage, or grain. Molds inhibit the normal digestive processes and lessen feed intake or can even cause illness in your cow.

not low enough to prevent mold development. The presence of mold indicates the hay wasn't allowed to dry enough before it was baled. Molds can cause serious respiratory or digestive illnesses, although the white molds most often found in hay generally make it more unpleasant to eat rather than causing illness. The same holds true for corn silage. Do not buy or feed your cow any corn or corn silage that contains a pinkish mold. Mycotoxins can have deadly consequences if eaten in sufficient quantities.

Some large square balers contain an attachment to spray small amounts of proprionic acid on the loose hay as it is funneled through the baling chamber before compaction. This is a way to depress mold development. Ask if this has been done with any large square bales you may purchase. Sometimes these amounts get too high and the salts deter cows from readily eating it. To be safe, don't purchase hay that has been harvested this way.

If you open a large square bale and see a browning or caramelized color in the center, this is an indication that heating has occurred inside the bale because of too much moisture. This browning will have a negative effect on the quality and palatability of the hay. Pass on it, and find something better for your milk cow.

Poisonous Plants

Several plant species are toxic to cattle. If there have been cattle on your farm previously, it's likely that few, if any, poisonous plants exist in the pastures. However, watch for them in areas you plan to use that haven't been pastured before.

Poisonous plant species vary by region. Larkspur, hemlock, buttercup, and nightshade are some of the more common. There are dozens of other plant species that can lead to poisoning, but learning about and identifying any that may be in your area will help you remove them from your property. Check your pastures or areas where you plan to graze your cow and look for poisonous plants. Some grow in wet areas or along a ditch or stream. Generally one plant will not affect your cow enough to cause major problems, but sustained access to them might.

More About Feeding Practices

There are many nuances to feeding programs. This chapter provides a general overview of what your cow requires for feed. The information serves as a basis for further study. Feeding a cow can be straightforward when there is an abundance of grass in your pasture. Or, it can be a challenge during extremely dry years when pastures produce little and the feed situation in your area becomes stressed and prices rise. In such an extreme case, you may simply dry off your cow and wait until pasture conditions improve or walk her along the roadsides to eat. It may mean you have little or no milk to drink, but you will still have your cow.

Managing Manure
Just Compost It

■   ■   ■

M A N U R E happens, and it adds up. It's the natural byproduct of your cow's digestive processes. Handling it in the most effective way is called nutrient management. Ultimately, manure is a valuable asset that can be used in a composting program.

Nutrient management programs are required when large numbers of animals are present on one farm. Even though you may have only one milk cow, you can also benefit from a plan that uses manure to improve your farm's soils or gardens. It is important to minimize or eliminate any detrimental effects the manure produced on your farm may have on the surrounding environment, particularly waterways, streams, or other water sources, such as wells.

Many problems encountered with manure result from runoff of the highly concentrated nutrients. This occurs when the manure is spread on sloping hillsides or in areas where water runs after heavy rains but is dry at other times. Most often runoff occurs in connection with frozen ground where the manure has little or no chance of being absorbed into the ground prior to rain or snowmelt. Farm waste runoff into streams, creeks, rivers, lakes, and other water bodies has become one of the most contentious issues between rural landowners and urban populations that view these waterways as wildlife and recreational areas. This issue, however, does not have to become a problem, and the manure from your cow can be an asset instead of a liability.

How Much and How to Use It

A mature dairy cow weighing 1,200 pounds produces 100 to 115 pounds of feces and urine per day, or 18 to 22 tons per year. Add to this the straw, sawdust, cornstalks, wood shavings, dry hay, or whatever else is used for bedding, and it can amount to a sizable quantity of material for disposal.

Left: Compost piles are an excellent and environmentally sound method of handling manure. Composting turns the manure into neutralized fertilizer for garden use.

Cattle will seek water to cool themselves during hot weather, but this can cause pollution of waterways. Building buffer strips or fences prevents them from accessing the stream and prevents pollution. *Petar Neychev, Dreamstime.com*

Manure quickly decomposes under warm, moist soil conditions and releases nitrogen, phosphorus, potassium, and other nutrients into the soil. Field plants absorb these nutrients from the soil in order to grow. Replenishing these nutrients with manure lowers or eliminates the need for outside fertilizer purchases. Besides helping with fertility, manure is used to improve soil structure and composition.

If pastured instead of kept inside, your cow can spread half or more of her yearly manure production by herself. Spreading cow manure pats across the field can be a good thing. The fresh pat slowly dries in the sunshine, and its thick

A fresh manure pat will slowly dry in the sunshine. Its thick consistency does not make it vulnerable to runoff from pastures like the liquid manure typically stored in pits or lagoons on large farms.

Manure pats decompose quickly when they are spread around the field. They provide nutrients for plants and homes for insects that feed on the fibers and organic matter.

consistency does not make it as vulnerable to runoff. Manure pats are a fundamental part of the pasture ecosystem. As a cow moves about, she deposits manure in different places. Her pats slowly decompose and provide nutrients for plants and homes for insects that feed on the organic materials. These insects in turn become food for grassland birds.

Manure is good for your soil. Manure, when mixed with bedding such as straw or hay, can add fiber and organic matter back to the soil. This combination, when incorporated back into a field or garden, can loosen soil particles and allow the soil to become more porous, absorb more moisture, and create more capacity for holding water. Loosening the soil particles relieves compaction of the soil. This creates a soil structure that helps plants develop a better root system for better growth.

Compost

The high nitrogen level in manure can burn plants when applied fresh directly to your field or garden. This is where composting comes in. Composting is the process of decaying and converting organic matter from a volatile compound to a stable, neutralized fertilizer.

You might already have experience composting grass clippings, leaves, and kitchen scraps. Now that you have a cow, you can compost her manure too. It's one way to handle it in an environmentally sound manner. It also can be a potential revenue source for your farm. Microbes and oxygen are the main catalyst for the composting process. The organic matter (solid manure) is allowed to decay in a pile or windrow. As it decays, it creates heat that breaks down the organic compounds and neutralizes them.

Composting manure can create a lot of heat. Depending on its density, the compost pile can reach core temperatures of more than 160 degrees Fahrenheit (71 degrees Celsius). Since oxygen is needed for composting, turn or stir the pile every 7 to 10 days. This allows oxygen to reach the core material and allows the edges to become part of the internal heating process. A tractor with a front-end loader or a skid-steer loader can easily accomplish this task for a large pile.

Wire can be used to build a compost bin. Several posts hold up the wire. You can fill the bin with manure, kitchen waste, and other organic materials that decompose.

The size of your compost bin will depend on the storage space available and the number of animals on your farm. A compost bin does not need to be an elaborate structure to serve its purpose. *Claudius Thiriet, Photolibrary*

The composting process doesn't happen overnight. To reach a neutralized state may take anywhere from three weeks to eight months, depending on temperature, the moisture content of the pile, and how closely it is managed.

Cow manure is among the best to compost because its nutrients have already been partially stabilized in the long digestive process of the animal. It contains about 3 percent nitrogen, 2 percent phosphorus, and 1 percent potassium.

Composting has many advantages. It transforms manure into a more stable nutrient form. It reduces the volume of manure you need to dispose. It allows you to store manure until weather or field conditions are better for hauling. It adds organic matter to the soil and improves the water-holding capacity and aeration of the soil. When spread on the field, these nutrients are slowly released into the soil for crop nourishment. This results in less nutrients being transported off the field through runoff or leaching into the ground water.

Composting has other advantages. It reduces the prevalence of flies by eliminating their breeding ground. It reduces parasite re-infestation since the heat generated kills parasite eggs. It neutralizes ammonia gases and certain pathogens such as *E. coli*. And it kills weed seeds, as they cannot survive the heating process.

Handling manure does not need to be a complicated process. The simpler, the better. A four-wheeler pulling a manure cart is a handy way to move manure on a small farm.

Even small amounts of manure, such as pictured here, should be piled on a solid cement surface or plastic sheeting until it is hauled to the field. This minimizes runoff and leaching.

Managing the Composting Process

How often will you clean out the barn, stable, or paddocks? It depends on your time availability, weather conditions, and the amount of effort you want to put into it. Frequent cleaning keeps the stable fresh but also increases the work load. You probably want to clean out the barn at least twice a week to avoid having to face really huge cleaning projects and to reduce fly populations in the summer. Fortunately, you can store manure and spread it on fields to suit your schedule. You can also store it by composting. Both allow for a more convenient schedule than daily or weekly spreading.

You can compost in windrow piles outside or construct simple compost bins. Use 8-foot landscape timbers, cement blocks, or other solid materials that withstand acidic conditions. Three bins with 8-foot sides that are 3 feet high can hold about 200 cubic feet of compost. Having one open side allows you to enter and easily deposit the manure and provide enough room to turn or stir it and also remove it later.

Plan a series of these bins if you have several large animals on your farm. One bin can store daily waste, another bin can be filled and left in the composting stage, and the third bin can have finished compost to be stored for when you need it or sell it.

Although bins are the easiest way to store the manure, you can also stockpile it on a level area where runoff is unlikely. Put the compost pile outside in the open, or use wood pallets placed on end and nailed together to hold the manure stack.

Whichever method you choose, avoid areas near water or surface areas prone to runoff. Stay away from hillsides, low spots, waterways, streams, ponds, or household wells, particularly if your farm has light porous soils.

Manage the composting process. This may involve tarping, turning, piping, and watering. The microorganisms require air and water to do their work properly, and they may need your assistance.

Using a tarpaulin, canvas, or other form of heavy covering prevents the manure piles from becoming soggy in the winter or too dried out in the summer. This covering helps prevent runoff of nutrients too.

Turning the manure pile allows oxygen to get to the bacteria and microorganisms that break down the manure into a soil-like substance, and it creates a more uniform exposure to the pile's heating process. Air will permeate the pile on its own to some extent, generally to a

A manure fork is a necessary investment with a family cow.

A skid-steer loader can be used to turn your compost pile. It is an easy way to stir the outer portions into the middle and keep the heating process going.

depth of 2 to 3 feet. It will be less if the pile is composed of solid manure with little bedding mixed in, and more if there is more bedding and less solid manure.

The method you use determines how often you turn the pile. Mechanical means, such as a tractor with a front-end loader or a skid-steer loader, make for fast, easy work. Another option is to use a manure fork and your strong back. In either case, turning the pile every 7 to 10 days will aid the composting process.

You can reduce turning work by introducing oxygen into the center of the compost heap via an **air-piping method**. Do this by using 4-inch-diameter PVC pipes with 1-inch pre-drilled holes spaced 4 inches apart. Lay these against the sides of your compost bin at an angle and add the manure and bedding around them as you fill the bin. These can be easily removed later when you clean out the pile. You can also insert pipes into an existing pile. Be sure to seal or tape the open end before you push the pipe into the pile to prevent the pipe from becoming filled with manure.

A tractor and spreader are needed for handling large quantities of manure. Spreading manure on fields is easier and more evenly distributed this way.

Even if you use a piping method, you will still need to turn the pile at some point. The outside of the pile needs to be turned into the center so the core heating process can kill parasites and weed seeds. This will also help you achieve a more complete composting, which is particularly important if you are planning to sell it.

The manure pile should have the consistency of a moist sponge. If it doesn't, you won't get the needed heating reaction between the water and the fibrous materials. Bedding materials such as wood shavings, corn fodder, and straw are very absorbent. That's why they are good to use. Putting these drier materials onto a compost pile may require additional water to saturate them.

Rain- or snow-soaked manure and bedding will require little added water. But if your pile becomes too dry, you can **moisten the pile** every time you add a wheelbarrow of manure and bedding, or when you turn it. The piping method has the added advantage in that you can flush in a short burst of water down the pipes occasionally. You can also use a lawn sprinkler to moisten the top of the pile if it becomes too dry.

Using these simple guidelines provides you with ready-to-use compost in as early as three weeks. It may take up to eight months, however, for a finished compost product if it does not heat properly because it isn't moist enough or if it hasn't been turned enough.

Once your compost is ready, you can apply it to your garden or flower bed. It will improve the health of your soil and plants regardless of where you live. It is one of the many advantages of having a family cow.

Milking

Adjusting Your Cow and Yourself to the Process

■　　■　　■

A FAMILY COW can bring great personal satisfaction and enjoyment. Just having her around and attending to her can become a daily pleasure.

Yes, your cow needs daily attention, and you must provide the basic care to sustain her and promote growth and good health. Still, cows are remarkable creatures. Left to their own devices, they can live fairly well on their own with little outside help. Their survival instincts can be to your advantage. In normal circumstances and with a healthy animal, you won't need to babysit her too often. After providing her with water, shelter, and whatever feed she can graze, she can be left alone for most of the day.

In return, she'll reward you with her milk, often considered nature's most perfect food. That's not a bad trade.

Making Milk Starts with a Calf

The initiation of lactation, or milking, begins with your cow giving birth. Before this happens, you'll need to make several decisions. If you buy a pregnant cow that is dry (not milking), your process will be quite different than if you purchase a cow that is in lactation (giving milk).

If she is dry, which I recommend, you will have some time to get her acclimated to your farm and you can plan your milking routine. If she is giving milk, you'll need to put your milking plan into action when she arrives.

A dairy cow is typically milked twice each day, seven days a week, until she is dried off in anticipation of her next calf. Peak milk production occurs in the first three to four months after calving. It is a standard lactation curve: high early on

Left: Before the introduction of machines, milking was done by hand. Life was slower paced then. There is a renewed interest in this practice as more people return to a rural lifestyle.

and then tapering off. Feed quality and availability, weather and climate conditions, and udder and body health are factors that affect production levels. There is no rule or law that says you have to milk her twice a day, once a day, or even at all. Nor is there any regulation that states what hour of the day she should be milked. These are all decisions that you make.

Whether you have one cow or more, there are three basic options for the milk produced: drink it, sell it, or turn it into another product. What you do with your cow's milk determines how many times a day you milk her. The more milk you need, the more often you milk her. It may not be three times a day, but twice a day is a good option for many families. If you plan to use a lot of milk, whether for drinking, for cheese-making, for butter-making, or for making other products, you'll need more milk. In that case, keep your cow in full production as long as you can. If you need less milk than planned, you can reduce the number of milkings each day with no adverse consequences.

Dairy cows have the ability to adjust their production to the demand for it. Taking nature's view in the wild, if the calf required more milk, the cow would increase her production. If the calf needed less milk, she would instinctively lower the amount she produced because it wasn't required. This instinct has stayed with the domesticated cow.

The udder tissue acts like a sensor to the pressures produced by the volume of milk in the udder. A long period of constant pressure where little or no milk is removed is a signal to the tissue to decrease production.

Set a Daily Milking Schedule

Will you milk your cow once a day? Twice a day? How will this fit into your work or family schedule? Your cow is adaptable. You can milk her morning, noon, or night. She can adjust to a 12-hour cycle, or a 16-hour and 8-hour milking schedule. Cows are creatures of habit and will adapt to the hours you

The Milking Schedule

Some dairy farms milk each cow three times a day, every eight hours, instead of the more customary twice a day. They do this for several reasons. Milking a high-producing cow more frequently reduces the stress on the udder tissue due to the large amount of milk held in the udder. Relieving this pressure reduces the chances of udder damage, which may lead to infections. Milking three times a day can increase the total daily volume by as much as 15 percent. It makes more efficient use of labor and facilities too. However, you are unlikely to milk your cow more than twice a day and maybe only once a day, depending on your circumstances. Your family cow's production is not as rigorous as a large working dairy.

The production of milk in your cow starts with a calf.

maintain for milking, whether noon and midnight, five o'clock in the morning and five o'clock at night, or something in between.

Decreasing the number of milkings per day causes a reduction in milk yield. Skipping one milking a week reduces total yield by about 5 to 10 percent. Milking once daily cuts the yield by half in first-calf heifers and by 40 percent in older cows.

If you raise your cow's calf, it could free up your schedule. If you need to be gone for a few days, with no one able to milk your cow, the calf will help you out. Just give your cow access to hay or pasture and let the calf nurse until you get home and resume your milking schedule.

Milking Your Cow

Learning to milk a cow is like learning to ride a bicycle. It may be a struggle and feel awkward the first few times. But once you learn, you'll wonder why you ever thought it was hard. In fact, milking is easy enough to learn that other members of your family might like to get in on it. Everyone in your family can learn and take turns in the rotation. You may be surprised at how quickly your cow adapts to each individual who milks her.

Starting with a Dry Cow

I suggest buying your first cow when she is dry. She could be due to calve in a few days or in several months. Either way, a dry cow will allow you to train her to accept the pen or stall where she'll be milked and to become accustomed to you.

Have a bedded pen ready with hay and water where your new cow can lie down after her trip to your farm. This will be a calming way for her to relax her first night at a new home. Most cows settle in quickly, particularly if handled gently and quietly.

Use her adaptability and desire for routine to help her find her milking place and be comfortable with it. Let her do this at least twice a day, at the times you plan to milk her. If you have a stanchion stall or head-lock in her pen, start to train her by having her put her head through. You can sprinkle some grain in front of her and let her decide to poke her head through to lock it. This will condition her to look for the sweet-tasting grains as a reward. She will adapt to this routine after a couple days of practice and even look forward to coming into her stall. This lets her settle into a familiar pattern that will make the days after calving less stressful for her and for you. She can adapt to close confinement during the milking if she hasn't had that experience before. By allowing her to become acquainted with a comfortable and calm atmosphere, you will also provide yourself with a pleasant area to work.

Some homesteaders like to hand-milk their cow in the pasture in order to enjoy the outdoor experience. Listening to the rhythm of milk stripping into

You control the milking schedule. You set the times that fit into your daily routine. Your cow will adapt to the hours you milk, and she is flexible enough to change if you do.

Many older dairymen have fond memories of bringing the cows home from pasture. Often this was their job as they returned from country school at the end of the day. You can create this time-treasured memory for your children with your family cow.

Whether you hand- or machine-milk, it is easier if you restrain your cow. Pasture milking is possible, but it is still best to tie her so she doesn't walk away from you.

a metal pail while watching the evening sunset or morning sunrise may be the dream you've held. This will work if your cow is tethered and very docile. But any quick moves or surprises by something in the pasture can result in spilled milk.

Bringing your cow onto your farm before she calves also allows you time to get her acclimated to her new surroundings, feed, pasture, stable, family members, cats, and, perhaps, an expressive family dog.

Work with your cow while she is tied up eating grain or hay. Use a soft hairbrush and move alongside and gently brush her back, rump, and sides. This will calm her and let her get used to your presence. Talk to her as you brush her or have a radio on playing at a soft or moderate level. The more she becomes accustomed to your presence, family members, or pets before her calving, the less likely she will be disturbed or agitated afterward—when you want to milk her.

While there are no guarantees, buying a cow that appears calm and docile is the best assurance that she will be that way once she arrives at your farm. Yet, even though she may be calm upon arrival, the birth of a calf can alter some cows' attitudes toward you or anyone else that may appear as a threat to her calf. You will need to be aware of her attitude toward you before or after calving and the first time you milk her.

Most cows are calm enough to share their calf with you and will express physical or vocal interest in what you are doing with her calf. On rare occasions, however, cows assert their protection in ways that are injurious to their handlers. Be calm, be alert, and be safe until you understand her attitude. Training your cow to be in your presence before she calves will go a long way in reassuring her of your intentions.

The first milking session after calving is a time of introduction to your methods. Your cow should go easily into the milking stall if she's been trained to do so beforehand. Making the decision to separate your cow from her newborn calf for the first milking depends on how much space is available. Three of you can occupy it at one time, if desired. The arrangement or location of the milking stall in relation to the calving area, and whether your cow seems attached to her calf, are factors too. Most importantly, keep yourself safe. Avoid spaces where you can become wedged against a wall or gate and be crushed if she moves from side to side. Give yourself room to move away from an ungrateful cow if she decides to kick at you. Close quarters make it more difficult to evade swinging legs directed at perceived pests.

By mid-lactation, your cow will understand the milking process. She'll generally accept her new surroundings with little fuss, especially if she has some feed to distract her. Her main concern may be how you milk her, whether by hand or machine.

Ways to Milk

There are two ways to milk a cow: by hand or by machine. Your idyllic view of hand-milking may be one of your reasons for buying a family cow. It can create a close bond between a cow and a human, but it also can seem an arduous task at the beginning.

Don't be discouraged. But realize that hand-milking a cow that gives forty pounds of milk a day, or twenty pounds each milking, is roughly five gallons. This will take some time out of your day to do a complete milk-out. It will test the strength of your arm and hand muscles. You may need to stop periodically during your first sessions to let your muscles relax, but practice will make you stronger. Over time, your muscles will increase in both strength and stamina. After the cramps in your hands and arms are a thing of the past, you'll look back and realize it is better exercise than joining a health club or mimicking the latest exercise video.

Hand-milking is easier if your cow has teats between two and four inches in length. Teats shorter than two inches are difficult to fully grasp with four fingers. While they work well with a milking machine, short teats only allow for your thumb and first two fingers of each hand to be used. This makes hand-milking more difficult than it should be. Teats that are more than four inches in length are not ideal, but they're at least better for hand-milking than machine-milking because of the design of many teat cup liners that are part of the milker. A median teat size will make full use of your fingers and hands during the squeezing motion needed.

You can become an expert hand-milker in a very short time. You'll develop a rhythm that suits you and your cow, and the daily milking may be time of peace and quiet. Call it meditation by milking. But whatever name you call it, it can have a therapeutic effect on daily stress. Experience is the best teacher, and you will quickly learn that long, slow strokes yield more milk in a shorter time than short, fast hand strokes.

If your time is at a premium or your physical ability is a major concern, a portable milking machine could be your answer. Maybe you're just not up to milking by hand. Regardless of the reasons, you don't need to feel guilty about using a machine. Portable milking units are available and affordable. They accomplish the milking in a very short time. They achieve a complete milk-out and are easy to clean and store until the next use.

Where hand-milking may take upwards an hour each time at first, a milking machine can accomplish the same result in five to eight minutes. This eliminates the physical stress on your arms and hands. If your cow is already accustomed to a milking unit being attached to her teats, you should experience few problems in milking her. Switching a machine-milked cow to hand-milking is more challenging than switching a hand-milked cow to a machine.

A milk machine unit operates most effectively when there are four working teat structures, although they can be successfully used on cows that have only three or even two healthy quarters. A machine works on the same principle as hand-milking in that it extracts milk from the udder. But it does so using the pull of a vacuum instead of the forced extraction used in hand-milking.

A milking machine can be used even for one cow. Each teat cluster and bucket is a self-contained unit. *Marcus Hasheider*

A milking machine uses a power source to drive a motor attached to a pump that creates a vacuum that travels along a hollow tube to the inside of the teat cup. The teat cups are composed of a rigid outer shell of stainless steel or plastic that holds a soft inner rubber liner called the inflation. The space between the outer shell and the inner liner is called the pulse chamber. A continuous vacuum alone would not milk the cow. It would only draw the milk to the bottom of the teat canal, where it would sit as well as drawing blood from the surrounding vessels to the teat ends. An electric pulsation device causes an intermittent break in the continuous tug, and this creates the squeezing and massaging of the teat structure that draws the milk out, much like hand-milking.

The machine works in two parts, the front two quarters and the rear quarters, and the electric pulsations alternate between the front and back. As the front quarters are being milked, the back quarters are at rest, and this alternation continues until she is finished.

The vacuum cycle creates a squeezing motion of the rubber teat cup liners, mimicking your hand squeezes against the teats. It also massages the teat structure, while the vacuum within the teat cup draws the milk from the teats while being squeezed. The milk flows into a cluster uniting all four teat cups and then through a tube directing the milk into a milk bucket.

There are many different makes and models of milking machines, but they all operate on the same principles. Talk with a local dairy equipment supplier to learn about the machines that may be available if you decide hand-milking isn't for you. The supplier can tell you about maintenance, availability, and service in order to keep your machine in good working condition. The beauty of it is that if the machine fails or the power goes out, you can go back to your original plan of hand-milking.

Keeping Her Udder Healthy

Your cow's udder is where fluid milk collects before it becomes available to you or her calf. Her udder is divided into four separate compartments referred to as quarters. Each consists of soft, pliable tissue made up of millions of minute pockets of alveoli that resemble micro-grape clusters. These clusters reside in the thousands of branches of the small milk ducts that are part of a larger milk duct system leading to the milk gland cistern. Almost all of the milk a cow is going to give at one milking is in the alveoli and ducts leading to the cisterns when milking begins. The alveoli have been using the mix of fermentation ingredients in the cow's bloodstream during the previous few hours to produce milk. Milk synthesis occurs in the epithelial cells lining the alveoli. Within the epithelial cells, milk fat is produced in small, well-defined droplets. These droplets merge as the move toward the central lumen of the alveolus and enlarge. Protein molecules are made the same way.

At a certain point, each alveoli becomes full and the pressure created then in each cell becomes so great that milk production ceases within it. This is why it is possible to get more milk from a cow through three milkings daily; fewer alveoli shut down and instead continually make more milk.

Each of the four quarters is separate, and there is no transfer of fluids between one to another through the membranes. This discreteness is one reason that three can continue to function even if an injury or infection, such as mastitis, destroys the production in the other one. There is, however, a difference in the amount of milk produced by the front and rear quarters. About 60 percent of the normal total milk yield will be produced by the two rear quarters.

Within each quarter and located toward the bottom is an area referred to as the annular fold; this leads to the teat cistern where the milk collects before entering the teat canal. Milk is held from exiting the teat by the sphincter muscle, which acts like a sealing band at the teat end.

Milk extraction from an udder is aided by the cow's hormonal response to stimulation of her udder, either by you or her calf, and regardless if milked by hand or machine.

Manual udder stimulation by applying warm water, a calf sucking, or hand massaging sends a signal to a cow's brain that initiates a release of oxytocin. This is a quick-acting but short-lived hormone that, once released in her pituitary gland, travels through her bloodstream. When oxytocin enters her udder, it causes the secretory tissues and alveoli cells to contract, squeezing out the fluid milk molecules trapped within them. These milk molecules enter the numerous pathways, each becoming larger as they near the teat canal, much like small roadways feeding into superhighways.

If allowed to nurse directly from the cow, the calf will only drink until it is full, not until the udder is empty. You must milk out the remainder yourself. *Shutterstock*

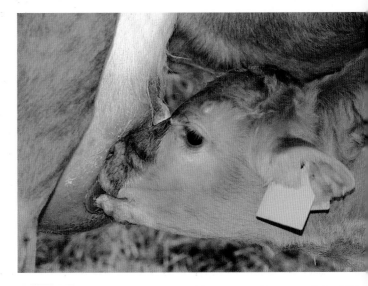

This movement of the milk into the teat canals is referred to as "milk let-down." The effect of oxytocin generally lasts about ten minutes. After that time, its effects become negligible. By then, however, the milk has been released from the cells and is waiting to get out of the teat end.

Milking machines typically do the best job in completely removing milk from a cow's udder. Hand-milking, if performed long enough, can accomplish the same thing. A calf will become full before all the milk has been removed from the udder. This presents a couple of options: you can finish by hand or by machine for your own use, or you can remove all the milk yourself and then feed the daily amount needed by the calf separately. Removing the calf from the cow soon after birth will require that you provide it with nutrition rather than the calf seeking its own from its mother. It's a choice whether you want to let the calf nurse or not. Do you need all the milk? If so, then you probably won't want the calf to have it.

Feeding the calf by hand several times a day will let you monitor its intake and observe it for any signs of illness. Bottle-feeding a calf has been done successfully for generations and is no less a humane practice than having it try to feed itself and not succeeding.

How to Milk by Hand

Hand-milking is a simple procedure. You will either become adept at it quickly, or you will soon get a milking machine. It is best to milk with short fingernails to reduce the risk of surprising your cow by accidentally digging into her teats. Also, remove any finger rings or wrist jewelry as these often get in the way of milking and are easily dirtied.

Let your cow know that you are beside her or are approaching so she's not startled or surprised by your sudden presence. Talk to her, pet her, rub her, and do everything you can to create a comfortable atmosphere at milking time.

Sit on a stool or plastic pail to one side of your cow. Don't sit behind the cow. The side approach is best for several reasons. Two teats are close to you, and you can lean against the cow's flank for support and discourage her from lifting her leg. Trying to come at a cow from behind her is awkward, and you can only reach the hind two teats. You then are left with the front two teats located on opposite sides still to be addressed. Finally, a cow's kicking motion involves leg movement slightly to the side before she straightens her leg out in full force toward the back. This happens in a split second. If you do something that causes her to kick, you will likely receive less of an impact if sitting on the side of the cow than in back of her. A direct blow can lead to serious injury. A full-force kick in the face by a cow's hard hoof likely means a trip to the dentist, hospital, or both, if not worse.

Once you are situated on your stool, it's essential to clean the teats before you begin milking. This ensures milk let down, but even more importantly it prevents udder infection. Mix warm water with a disinfectant such as a mild dishwashing liquid, household chlorine bleach, or similar product. Dip a clean cloth rag or a sturdy paper towel into the water and gently clean around the sides of each teat to remove all mud, dirt, manure, or other foreign matter. Pay close attention to cleaning the end of each teat thoroughly. Rub the sides and back of the udder to help stimulate milk let down. When all the dirt has been removed from the teat sides and ends, use a fresh cloth or paper towel to dry them off. Proper sanitation is the best insurance against introducing microorganisms to the teat canal. Very quickly the udder will become tighter as the released milk starts to collect in the gland cisterns.

Proper hand-milking technique is to hold the top part of the teat near where it attaches to the udder. Hold the teat firmly between your thumb and first finger and then squeeze them together. This traps the milk within the teat canal and prevents its escape back up into the udder cistern. Then bring your other fingers

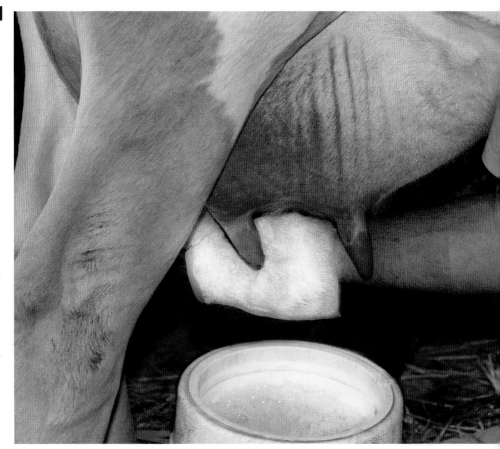

Before milking, always wash the udder and teats with a mild liquid detergent or dairy soap and warm water. This is good sanitation and will stimulate a milk let down response in your cow.

Udder washing products are available from dairy supply companies or farm supply stores. These are often diluted in water. They provide an effective cleaning and sanitizing effect.

Paper towels or cloth rags are used to clean the udder and teats before milking. Dispose of paper towels after each use. Do not use sponges, as they can harbor bacteria regardless how clean you may think they are.

together, one at a time, in a downward direction, squeezing them together and forcing the milk out of the teat end and toward the pail.

A proper technique involves squeezing the milk out rather than pulling it out. Pulling constantly on the teats will irritate your cow and you'll likely receive one of those sideways foot messages.

When you first practice hand-milking, you may need to milk a single teat at a time. Ultimately, your goal is to milk two teats at a time, in alternating strokes. After squeezing the milk out, release your grip and let the teat canal refill with milk. This should happen instantly. Once you have mastered milking two teats at a time, you can alternate each hand motion to develop a rhythm until you've emptied two quarters.

Then move to the other side of the cow and repeat the milking process for those two quarters. Unless your cow has gotten dirt, mud, or manure on her udder while you were working one side, you won't need to rewash the unfinished teats before you start.

Step 1. Hand-milking is a simple procedure. Start by gripping the teat high, near where it adjoins the udder, with your thumb and index finger to form a circle around it.

Step 2. Close off the top of the teat by squeezing your thumb and finger together. Then squeeze the rest of your fingers in a downward direction.

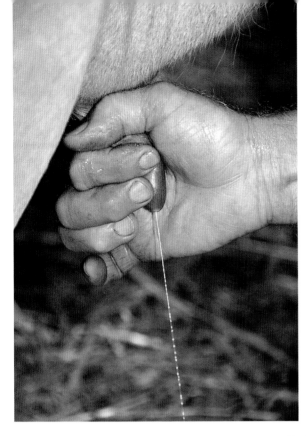

Step 3. Close your fist so the milk is forced down through the bottom of the teat and out the opening. Then open your hand and let the teat refill with milk. Repeat motion.

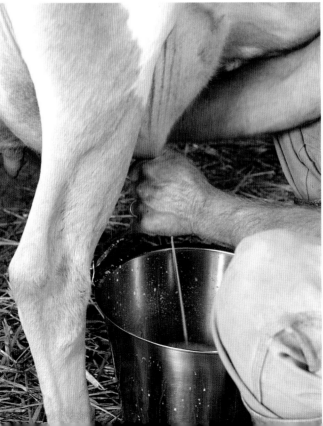

Eventually you will be able to milk two teats at a time.

The Milk Pail

Choose milk pails that are made of easy-to-clean material, such as stainless steel or plastic. All pails must be thoroughly cleaned between each use. If they're metal, be sure they don't have any cracks or rust spots that could harbor bacteria.

Partway through milking, you may want to dump the milk obtained so far into a second pail sitting a ways away from you and the cow. It's good protection if she tips the pail over with her foot or you accidentally tip it yourself and lose your efforts. No crying over spilt milk. Hand-milked cows are known to shift their weight and plant a foot in the milk pail, which would render the milk undrinkable.

Using a Milking Machine

With machine-milking, you attach all four teat cups at one time instead of doing one side and then the other. Be sure to thoroughly wash the milking equipment after each milking and sanitize it before it is placed in a clean area where it can air dry. The milk cluster and teat cups should be sanitized after each milking and again rinsed in a mild disinfectant and then clean water before they are used at the next milking.

When you have finished milking and all four quarters are completely emptied, dip the teat ends in an iodine solution or another approved dairy disinfectant blocking agent. This will kill bacteria that may attach to the teat end during milking.

Apply iodine or other antibacterial solutions to the teat ends after milking is completed. This will seal and sanitize the teat opening and prevent bacteria from infecting the udder.

Bacteria cells grow and multiply, and they can grow through the teat opening normally protected by the sphincter muscle. This is a very strong muscle to hold milk within the teat and udder, but it's not tight enough to prevent bacterial growth. Good sanitation pre- and post-milking is your best insurance for keeping your cow's udder healthy.

Colostrum Milk for the Calf

A fresh cow's first milk is yellow-colored and has a thick, sticky consistency like cream. This is called colostrum. It is nature's protection for a calf and contains a rich mixture of proteins, vitamins, and antibodies. When these are absorbed in the calf's fragile digestive system, it helps protect the calf from colds and infections in the first days of its life.

The calf's digestive system can only absorb the colostrum for twenty-four to thirty-six hours before its usefulness decreases. It is important to feed this to a newborn calf to help it transition from the birth process.

The production of colostrum coincides with the approach of calving. Freshening cows have udders that expand and fill with fluids in anticipation of having a calf to feed. This is a normal physical response.

Some cows will experience increased tenderness in their udder tissues at the time of calving and may be more sensitive to your touch as you begin to milk them. The swelling and redness of her udder diminishes during the first week as the colostrum begins to switch to normal fluid milk.

By about the sixth milking, the yellowish colostrum will have turned into a normal white fluid. Your cow will produce more colostrum than her calf can use. What do you do with it? Save it by freezing it in plastic calf nursing bottles available at many farm supply stores. Freezing colostrum does not harm the proteins or antibodies so it can be used later. Excess colostrum can also be fed to pigs, chickens, dogs, cats, and other small livestock.

You can also use colostrum for family recipes. Cooking will remove its bitter taste and will cause it to thicken. It can then be used in baking and for non-egg custards.

There is one caution with using colostrum for human food. If you bought a dry cow, find out if she's been treated with antibiotics. Many farmers routinely infuse antibiotics into each quarter of a cow before they dry her off. This is to combat possible infections during the time she stands dry. People who have allergic reactions to antibiotics or just don't want them in their diets shouldn't use colostrum from a treated cow. Most antibiotic residues are eliminated from her system during this period, but residual traces may appear in the colostrum. If she had been treated during her dry period, she will likely have eliminated all traces by the fourth or fifth milking after freshening.

Safe Milking Practices

Udder sensitivity may cause some problems for you during milking, especially right after calving. Your cow may injure her udder during her lactation and have a reluctance to let you milk her. She might kick at you because it hurts.

There are several ways to mediate this behavior. The easiest is using an anti-kicking device, which can be purchased in a variety of styles. Many are priced reasonably. One of the easiest to use is a flank-style restraint that is shaped like an upside down letter U. It has a screw handle at the top that tightens the sides against her flank. It is placed between a cow's hips and thurl joints at the pelvis from the top of the cow down. When the hand screw is tightened, the long metal bars press against her flank and immobilize any forward or sideways movement of her leg. It won't hurt the cow or cause pain because it's not a sharp object. But it will restrain her legs so that she can't lift them to kick. This device is one of the most humane cow restraints available to prevent kicking.

Another alternative is to use a rope tied to one front leg, between the knee and ankle and pull it so her foot is off the ground. The rope can be tied to a post or over her shoulder and tied to her other front leg. Her automatic response will be to balance herself on the other three legs. It is virtually impossible for a cow standing on only one front leg to kick you. This is also a humane method of restraint, but it takes more effort on your part to raise her leg. If you use this method, make sure your cow is head-locked and can't get loose. Chasing a three-legged cow hobbling down the cow path won't make for a pleasant morning experience.

A third option for restraint is to grasp the tail near the vulva and push it up into the air at a 90-degree angle. The pressure created against her rear spinal column will immobilize both rear legs. This maneuver is sometimes used for quick action restraint, such as injections, and generally requires a second person to help. The biggest disadvantage is that the one holding this posture can't do it for very long before their arms tire and the tail is released.

With whatever form of restraint used, two important considerations are safety and patience. No form of restraint should ever be used that will cause your cow injury or pain, such as chains, cables, wire, or other metals that can cut into her legs or body. Pain will cause an immediate response from her, which may be turned against you. Patience is said to be a virtue; with your cow, it is an imperative. Losing your patience and hitting her will quickly undo all the good things you've used to put her at ease. She's likely to strike back. Be patient. Be gentle. Be humane. She'll respond better to you if you are.

Right: A hip restraint is a humane way to keep a cow from kicking during milking. It provides a measure of safety for you.

Udder Treatment

You will soon develop an expertise in milking you didn't think possible at the beginning. As you become familiar with the process, you will recognize any changes in the udder, such as hardness of tissues, that may indicate an infection resulting in mastitis.

You can easily check for infection at the beginning of each milking by stripping the milk from each teat into the palm of your hand. Healthy milk will look like normal milk while mastitis milk will look like clumps of cottage cheese in a watery solution. The clumps are masses of white cells that combat infections and indicate the presence of bacteria in the udder tissue.

There are several ways to handle treatment of mastitis, depending on its severity and your preferences. For a mild case, you can strip the affected quarter at five to ten minute intervals for several hours. This action continually removes the toxins produced by the bacteria so they do not have time to build up in quantities in the quarter. You won't get a lot of milk each time you strip the quarter out, but it does aid the circulation of blood in her udder, which carries the white cells to help the cow combat the infection.

A mild case can also be treated effectively with over-the-counter antibiotics, but this requires a specific withholding time for all the milk intended for human use and not just from the infected quarter. Mastitis milk can be fed to other animals with no ill effects.

Severe mastitis cases may require veterinary assistance and stronger antibiotics than is available over-the-counter. Extreme cases, such as coliform mastitis, require quick and immediate intervention because these turn systemic very quickly and often lead to death if not caught in time.

Teaching Her to Milk

Breaking a cow to milk is not a difficult process, but it does take patience. That may be reason enough not to buy a springing heifer that has not had a calf or been milked before. In this case, she will likely need training for being tied up or restrained as well as milking, which she probably won't understand. It's not impossible to teach new first-calf heifers about being milked. It's done all the time. But it is more difficult than working with a cow that has been milked previously.

To restrain a fresh heifer, lock her head in a stanchion or stall with a head-lock and use a flank restraint. This is the easiest and safest way for her to understand the milking process. Knowing that she is being handled and milked but can't kick at you will help her settle in to your routine. As her udder pressure increases with milk production and the subsequent relief she received from the milk being removed, she may be more willing to let you handle her, especially if she receives a treat of grain at the same time.

The reward for your milking efforts is fresh milk. Use a stainless-steel or plastic pail to transport the milk. Don't use any container that it is rusty, damaged, or can't be thoroughly cleaned.

Keeping Milk Records

The principles of hand- and machine-milking are identical: to extract the milk efficiently and completely. How you choose to accomplish this with your family cow depends on your preferences and abilities.

You may want to record the milk weight at each milking to get an accurate idea of your cow's capabilities. A simple hanging material scale should be sufficient to weigh the milk. Record the weights. You may be surprised at your totals as each week passes.

CHAPTER 8

Calving Time
The Cycle of Life

■　■　■

LIKE A RUNNER starting off at a pistol shot at a race, once a cow's calf is born, her body systems shift into action. The process of giving birth triggers the initiation of lactation and milk production that can last a full year or more. Milk starts as colostrum produced for her calf; this lasts for a few days and then switches to regular milk. Increased milk production makes demands on body fat reserves that she's been storing during her dry period. She needs more nutrients to meet that demand. She requires more water.

The cycle of calving and milk production happens because a cow was impregnated either by using artificial insemination (AI) service or by a bull with natural service. It is difficult for any cow to produce milk for several years without the benefit of birthing a calf. A calf every year keeps initiating milk flow. It's also nature's way for a cow to reproduce herself and continue the species.

Most well-fed cows that have had a six- to eight-week rest from milking will have their calf normally. On occasion there may be need to assist with the birth, and you will learn how to handle these situations.

This cycle of life involves three basic stages: calving (giving birth), breeding (insemination), and pregnancy. If you purchase a pregnant cow in mid-lactation, she will soon be ready to dry off from milking. If you select one that is already dry, she'll be getting ready to calve.

Your cow's dry period is a good time to get your facilities in order and prepare yourself for the calving and milking that follow. Have a place ready for the calf. If she's new to your farm, train your cow to use the stall where she'll be milked. This familiarity will ease her transition after she has her calf. If your cow has access to a dry pen, lot, or pasture, nature will often do the rest. A clean, dry calving area is ideal because it offers the mother and newborn calf freedom from mud and dirt. A mud-free birthing area lowers the risk of vaginal and cervical infections

Left: The birth of a calf initiates the milk-producing process. Good mothering instincts are important if you plan to keep the calf with your cow.

after calving. It also makes for a more pleasant work area in case you need to offer assistance to the birthing cow.

Dry pastures are a good place for a cow to calve because it is soft and has stable footing. But it generally does not allow for close observation or restriction of movement if help is needed. A calving area or pen allows you to keep an eye on the whole birthing process. You may have taken care in choosing a breed with smaller birth weights. You may have done everything possible to give the cow all that is required to make calving easy. But things can go wrong at calving time even though the best precautions have been taken.

Most of the problems that occur during the birthing process are caused by calves that do not exit the birth canal correctly, sometimes due to abnormal presentations. For various reasons, the birth may become difficult, and having the cow close at hand in her pen will help in correcting the situation sooner and easier.

What to Do at Calving Time

A cow's gestation period typically ranges between 277 and 283 days but can be as much as 260 to 296 days. If you know the exact breeding date, a good rule of thumb is nine months and nine days (279 days) later. There are also signs you can look for in a pregnant cow to indicate that calving time is approaching. The rapid expansion of a cow's udder or the full development of an udder in a young heifer is an early sign that calving is getting close.

There are four basic stages in any normal calving: pre-partum, labor, birth, and placenta expulsion. Stage one can last as little as three hours, although heifers may go as long as three days. During this stage, the broad ligaments on either side of the tailhead appear to sink away or relax. The cervix, vagina, and vulva are dilating to allow for the passage of the calf.

Stage two occurs with the opening of the cervix. There may be a long, stringy, and cloudy discharge. This is an indication that the mucous that sealed the entrance to the placenta is being expelled. Soon the placenta or the water bag appears at the vulva. As the uterus contracts and the cow strains, the pressure breaks this bag and discharges a great amount of placenta fluid. At this point, the cow may be up and down numerous times and start or stop pushing. By this point, the calf normally has started to enter the birth canal. Now you should check her every ten minutes to make sure the labor is progressing to the third stage, where the calf appears. Stage two can last from twenty to thirty minutes in cows and up to one to two hours in heifers. If it takes longer for the front feet to appear at the vulva, she may be having trouble with a difficult delivery. In this case, you will have to check the cervix.

During a normal calving, a cow enters stage three when the front feet and nose of the calf appear at the vulva opening. As the uterus contracts, it pushes the calf out farther until the head and shoulders begin to emerge. At this point, assistance can be given if it looks like the calf is larger than expected or the cow is having trouble pushing her calf out. Gently tug on the forelegs of the calf with a rope or obstetric chains attached to one or both of the legs of the calf above the ankles. Pull only when the cow strains and pushes. Time your pulls with her movements. Once the feet, head, and shoulders emerge, the rest of the calf will usually follow fairly easily.

A critical time in the birthing process is in the moments before the hips of the calf emerge. Just prior to this, the umbilical cord is still attached and the mother is still assisting the calf in getting oxygen. Once the umbilical cord is torn, the calf is on its own and must begin breathing by itself in order to survive.

The calf should be trying to lift its head quite soon after delivery. The shock of hitting the ground or the cold air seems to jumpstart the calf's breathing. If it does not, you may have to check the calf's airway to clear any mucous. Sometimes slapping the calf on its side or pushing on its ribcage can get it to start breathing. If the airway doesn't seem to clear, you may need to lift the calf vertically by its hind legs to let mucous and fluid drain out. Sometimes sticking a piece of dry hay or straw up its nose will trigger a reflex in an unresponsive calf.

Opposite: A calf is born after about nine months and nine days of gestation. Most calves are born without any complications, but be alert to potential problems. A live calf born in your presence is a wondrous event.

Stage four of the birthing process is the passing of the placenta, or afterbirth. It should be expelled as a large, gray-colored mass with traces of red blood veins running through it. Your cow may try to eat the placenta. She's just obeying nature's age-old directive of removing traces of a recent birth so predators aren't attracted to decaying animal flesh. Some cows may not expel the placenta as quickly as others, and in some cases, not at all. If stage four lasts over twelve hours, it's referred to as a retained placenta. It doesn't hurt the cow, and it's usually caused by a metabolic mineral imbalance during her dry period. Don't pull to remove it. It's best left alone. Pulling on this retained placenta can rip the sensitive cotyledons and cause extensive uterine bleeding. The cotyledons attach and hold the placenta to the uterine wall. Nutrients pass through these from mother to calf.

If after twelve hours the cow appears healthy, is eating, and does not have a fever, it is best to cut off any placenta at a level at about the top of the udder. Leave the rest. It should be expelled about a week after calving as it slowly disintegrates in her uterus and sloughs off the cotyledons still holding it. However, if she stops eating, ceases milk production, or runs a fever, you should contact a veterinarian for treatment.

Shortly after giving birth, a cow will get up to search for her calf. Let her lick it dry, and watch the amazing bonding that is taking place between the two.

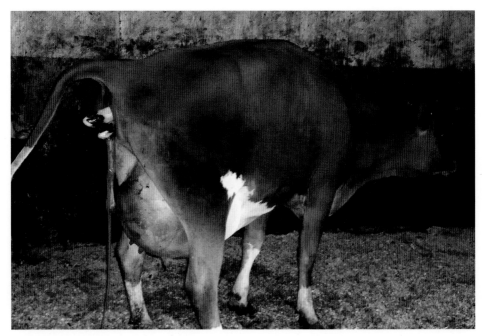

The three-stage calving process can move quickly. As the front feet appear, they should be accompanied by the calf's nose. If unsure of the presentation, you may have to reach in the cow to determine the calf's position. Don't assist your cow unless you decide it is really necessary. *Marcus Hasheider*

How to Handle Calving Problems

You can't predict when or how things may go wrong during a calving. But you can prepare yourself and increase your chances of a successful outcome—a live calf. Watch for two things during calving and post-calving that are cause for concern. Each requires a different approach and maybe veterinary assistance.

The most frequent problem during calving is an abnormal presentation. This is when the calf does not exit in the normal configuration of head and front feet first. Abnormal calf positions include:

- **Single leg flex in headfirst position.** You'll see one front foot is forward in the correct position, and the other is flexed back and inward.
- **Headfirst position with head and neck turned back.** The front feet are exposed but the head is turned back toward the inside of the cow.
- **Headfirst position with rear legs under the body.** The whole calf is trying to come all at once.
- **Hind feet first.** This is a backward calf. Many calves can be saved from a breech position with correct assistance.
- **Hock flex-hind feet first.** Calf is backward with the back legs turned downward.
- **Backward and upside down.** This is one of the most difficult situations to correct, but it's still possible to deliver a live calf.

Each of these situations requires immediate intervention. As the uterus continues to contract to expel the calf, it puts more pressure on the calf and the amount of room to correct these problems diminishes.

You will need to check the position of the calf before you know what maneuver to use in straightening it out. There are no x-rays for this; you need to reach in with your hand and arm and use leverage to turn the calf around.

Keep in mind four basic rules when a calf is not coming normally:

- **You can't usually change a posterior position (rear end first) to an anterior position (head first) because there's not enough room.**
- **You can't deliver a calf head first unless both legs are started through the cervix with it. All three parts must come through at once.**
- **You can't deliver the calf if three or more legs are through the cervix. The rear legs must be pushed back to proceed.**
- **You can't deliver the calf with only the head sticking out. It must be pushed back in and the front legs pulled forward.**

Before you go in with your hand and arm, thoroughly wash with soap and water and use plenty of lubrication, such as a mild liquid soap. You can use a rubber glove or obstetrical sleeve to cover your arm, but that, too, needs to be cleaned with soap and water.

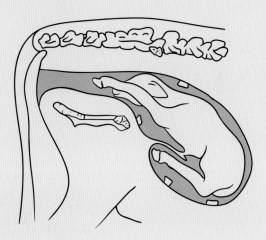

Normal position of a single calf.

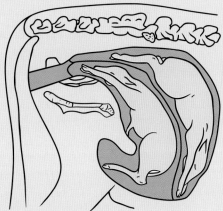

Normal position of twins.

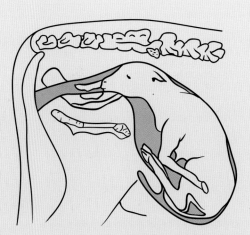

Single leg flex in headfirst position. One front foot is forward in the correct position, and the other is flexed back and inward.

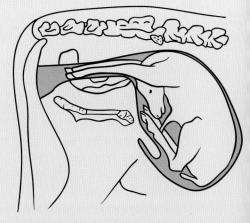

Headfirst position with head and neck turned back. The front feet are exposed but the head is turned back toward the inside of the cow.

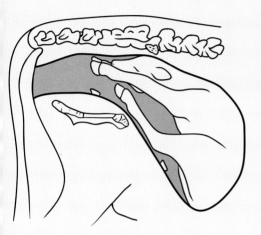

Headfirst position with rear legs under the body. The whole calf is trying to come all at once.

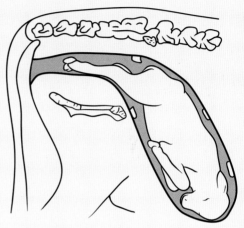

Hind feet first. This is a backward calf.

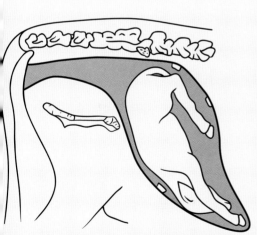

Hock flex-hind feet first. Calf is backward with the back legs turned downward.

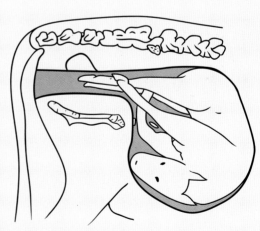

Backward and upside down.

You need to wash the vulva thoroughly before you gently insert your hand into the vagina. If you find the cervix is only dilated the width of two or three fingers, the cervix may still be in the process of dilation, and it may be a matter of waiting a little longer. If you don't think this is the case, it may be a twisted uterus. Call your veterinarian for help.

If you can reach in and feel the calf, then you can assess what position it is in and decide the best way to turn it for delivery. On rare occasions, it may be better to retrieve the hind feet and pull the calf out backward than to spend time trying to turn it all the way around so the front feet come first. It may be the calf has already been twisted around and any more twisting may strangle the umbilical cord, leaving the calf with no oxygen and causing death.

One of the more difficult birth presentations is twins, especially when they are tangled up. Usually one will present itself first and the other comes along quickly afterward. If you have enough time to call for assistance, that may be the best route. If time is too short, then you will have to determine which front feet go to which head and begin at that point. You can be successful in either case.

If you feel uncomfortable trying these techniques, you should contact your veterinarian as soon as you think there is a problem. But as you gain knowledge of what to expect, you should be able to handle most births by yourself.

What to Watch for During Calving

Torsion of the uterus is when the uterus becomes twisted inside the cow. The uterus is like a plastic bag filled with water, which during torsion becomes twisted until it is partially or completely closed. The common rotation of a torsion is about 180 degrees and will constrict the cervical opening. You can't pull a calf through this torsion; it must be straightened out first. Correcting torsion is difficult because you must turn a large calf lying deep inside the uterus with little room for your arm to maneuver it. All this is while the cow is restless and often straining to birth her calf. If not corrected, however, a uterine torsion can lead to death, both for the cow and calf.

There are two ways to correct torsion. If it is only a partial one, you may be able to grasp one or both front legs and turn them in the direction of the twist. As you turn, the calf and the uterus will follow along and straighten out. The calf can then come out. The second method, necessary in more severe torsions, is to use a long, stainless steel rod with a round loop on the end and an obstetrical chain to give you more leverage from outside the cow. The chain is attached to one or both front legs of the calf first and then the rod is inserted and the chain wrapped around the loop. The chain is wound around the rod as it is rotated in the direction of the torsion twist. When the chain is tight, the calf and uterus will rotate into the correct position. The chain, rod, and your arms need to be sanitized before insertion.

Both of these torsion corrections are a matter of mechanics, common sense, and hard work. Be clean and gentle because the uterine, vaginal, and cervical tissues are delicate, as well as the fetus. Be careful with your efforts, and if possible, get a veterinarian's help.

Uterine prolapse occurs when the cow strains and pushes her uterus out of her body after calving. It is an emergency situation, and you should call your veterinarian right away for help. It will be unmistakable because a large muscular sac will protrude or hang down from her vulva. If it happens outside in a pasture, slowly walk your cow back to her pen. Don't let her run because gravity will pull down on this large mass and extreme movement can rip and tear the blood vessels in it. If the uterine artery tears, the cow may bleed to death. Once in her pen, wash off any dirt or mud attached to the outside with soapy warm water. This will save time when the veterinarian arrives as he/she will need to move quickly. They will attempt to replace the uterus by pushing it back in and then suturing the vulva to keep it from coming back out. The majority of uterine prolapses can be successfully replaced, provided veterinary medical attention is given promptly. A medicine given to contract the uterus may prevent this from happening.

If your cow has had a uterine prolapse, consider finding a replacement later in her lactation. Cows that have had uterine prolapses can become pregnant like any other cow; once a cow has had a prolapse, however, the chances of it happening again are increased at the next calving.

What to Watch for After Calving

Milk fever, or hypocalcemia, can occur soon after calving, especially in older cows. This, too, is an emergency situation. It is not a "fever" where the cow's temperature rises but rather a lack of calcium in her body reserves. The process of calving and beginning the production of milk can cause such a huge demand for calcium that it depletes the calcium in the cow's blood supply to the point where she becomes weak and unable to stand. Because calcium is needed for muscle contractions, she may be wobbly, try to stand, and fall over. She may also have cold ears, which gave rise to the misnomer of a fever. Instead of rising, her temperature may drop. Milk fever is treated intravenously with calcium, which gives the quickest response, or possibly with an oral calcium gel. Left untreated, this is a lethal condition as the heart also needs calcium to function. In severe cases, it is the cow's heart that stops working before anything else.

Calving paralysis happens when the nerves in a cow's pelvis are injured during delivery. Nerve injury is caused by a small pelvic opening or a large calf that cannot pass easily. Either situation can exert pressure on the pelvic nerves, causing paralysis. Heifers are more prone to this condition than older cows because of a smaller pelvis. But older cows can have such huge calves that they experience it too.

Paralysis may last only a matter of minutes or several days. Signs range from a weak leg to not using a leg at all to the inability to stand entirely. Treatment can be made by steroid injections as well as injections of non-steroidal, anti-inflammatory drugs to reduce inflammation of the affected nerves. But generally, rest will do as much good as anything.

If your cow goes down because of calving paralysis or milk fever, it is important to have plenty of dry bedding underneath her. If a large cow lies on one side for more than three or four hours, the muscles on the down side can die because of the extreme pressure exerted by the cow's weight. You can mitigate the effects of this by turning the cow from side to side every one to two hours. You can push against her hip and gently roll her over onto her other leg. It is more important to exchange the rear leg positions than her front legs. Cows typically stand by rising on their rear legs first, followed by their front legs. This makes the rear leg movement essential.

Moving her from side to side aids the circulation in her leg muscles, which will give them a better chance to heal. Cows have been known to be down for ten days and make a full recovery. But it takes good nursing care for that to happen. If she is down for any length of time, you need to arrange a different feeding program for her calf. You can strip some milk from her udder to use. You can also begin to use a milk replacement product.

During the time she is down, she should have access to water and hay. If she makes an attempt to get up, you can help by grasping her tail at the base near the pin bones and lifting as she is trying to rise. Standing, even if it is only for one minute, is a psychological triumph for her as well as a physical one. She knows she can get up and will likely try again later after she lies back down. Standing, even briefly, helps with leg circulation and letting the blood get to all the muscles.

Don't let your cow out of her pen if she has had milk fever, calving paralysis, or difficulty getting up and down for any reason. At least not until she can prove she won't fall once she's outside the pen. It's easier to work with her in a bedded pen than on hard ground if she falls again.

What Comes Next?

Congratulations! You have a calf to raise. You are well on your way in milking your cow. Plus, you can start making dairy products from her milk for your family. What could be better? Well, how about another calf?

The next stage in your cow's reproductive cycle usually occurs forty-five to sixty days after calving. During the six weeks after calving, your cow's reproductive system (uterus, ovaries, and cervix) has been healing from the birth process and shrinking back to normal size. This period also lets her hormonal balance return from maintaining a pregnancy to preparation for another one.

A cow's ovaries lie within the pelvic cavity. They are similar to a bean in size and shape. Both ovaries are intimately related, working in conjunction, and are attached to the tissues that envelop most of the other reproductive organs. When a cow is "open" (not pregnant), the ovaries lie within or on the front edge of the pelvic cavity. In advanced pregnancy, the ovaries are carried forward and downward into the abdominal cavity along with the enlarged uterus containing the fetus.

The average cycle between estrus, or heat periods, is twenty-one days. Once your cow has begun cycling after calving, these periods will become very regular until she becomes pregnant.

The most obvious signs of a cow in heat are the standing posture of the animal and a clear mucous discharge from the vulva. If you only have one animal on the farm, it may be difficult to observe the posture sign. Other signs include not letting down her milk, constant bellowing, irritation, rubbing against you, or in extreme cases trying to jump on you. Her heat cycle may be difficult to detect because it may last only a couple of hours although it can range from twelve to eighteen hours.

Between forty-five to sixty days post-calving, your cow may exhibit estrus. The most obvious signs is your cow allowing another animal to mount her. If artificial service is planned, the cow should be breed within six hours of this sign. *Marcus Hasheider*

The Life Cycle of a Milk Cow

A new calf is born. If male, he is destined to become a stud bull, for breeding, or a castrated steer, brought to market weight and slaughtered for beef. If female, she may become your dairy cow, capable of producing a succession of calves and thousands of gallons of milk for you and your family.

A heifer is first bred as early as fifteen months of age in large breeds like the Holstein or as late as eighteen months in small breeds like the Dexter. Some nine months and nine days after a successful breeding, the heifer delivers her first calf. She also begins to produce milk for the first time in her life. Her peak milk production occurs during the next three to four months.

Some seventy to ninety days after calving, your cow is rebred. Her gestation period is again about nine months. In the final two months of gestation, she is dried off in anticipation of calving. This is her dry period, when her body is focused on nurturing the calf inside her and you may be able to take some time away the farm and the rigors of daily milking. When the calf is born, your cow freshens her milk production, and you again will see peak milk production for the next three to four months.

And, hopefully, the process repeats, year after year.

If other cattle are around, she may attempt to mount them, or she may be mounted herself, which is a standard indicator of a good heat. If you want a pregnancy from this cycle, you should make sure she is bred within six hours. If no other animals are around, you'll need to be a vigilant observer for the heat signals.

If you notice a slight blood discharge from her vulva, her heat cycle has passed and it is too late for conception this time. Note the date this occurs and mark your calendar nineteen days ahead, at which time you should begin to observe for her next heat.

Your goal should be one calf each year. This gives you the opportunity to milk your cow for ten months or about 305 days. This schedule gives her an annual sixty-day dry period before her next calf is born and may give you some time away as she will likely require little attention. Still, if you're away, have a friend or neighbor check on her daily just to be sure she hasn't gotten herself into trouble.

Maintaining a yearly calving schedule requires having your cow be pregnant seventy to ninety days after each calving. There may be some variation to this, but it is an achievable goal. Yearly calving yields the highest milk totals over several lactations and provides a reasonable balance of milking and recovery time for your cow in relation to the length of her gestation period.

There may be reasons to increase the length of her lactation and delay breeding her for longer than ninety days after calving. For example, you may want to have her dry so that you can be away at a given point during the year. Or, you many decide to milk more than one cow. Having more than one cow will allow you to have a more steady milk flow the entire year. When one cow is in mid-lactation and her productions starts to drop, you have a second cow freshen and come into full production. She then maintains her high level through your first cow's dry period.

If you plan to breed for annual calving, the ideal situation is to observe for several heats that exhibit a regular pattern and then breed your cow on the next heat that falls nearest ninety days after calving. If no observable heat period occurs within eighty days, then have her bred on the first one after that.

Artificial Insemination or Natural Service

You can use artificial insemination (AI) service or a bull for natural service to impregnate your cow. Keeping and maintaining a bull on your farm to breed one cow is not a recommended practice. Bulls are dangerous—to you and your family members.

AI involves using bull semen that is collected and processed under sanitary conditions at a bull stud. This is an organization that owns bulls from most every dairy breed and numerous beef breeds. There are many bull studs located across the country and around the world, giving you access to the best genetics available, with a range in prices for each service.

You can learn to do the AI yourself, but it may be simpler to contact an AI service technician in your area. You can contact any of the AI studs to learn of their nearest representative. Just as you prepared ahead to bring your cow to your farmstead, you should also have a discussion with the technician or stud representative well ahead of the time you want to breed your cow. They can explain the bull semen available, costs involved, and how and when to contact them as service is needed. If you have a cow of a rare breed, you may need to make special arrangements for semen from a bull of that particular breed. To use AI effectively, you will need to restrain your cow so the technician can deposit the semen into her uterus. Tying her in her milking stall will be sufficient.

If your cow doesn't come back in heat nineteen to twenty-two days after having been bred, the chances are good that she has become pregnant. If she repeats her heat cycle, you will need to call the technician for another service. Most AI technicians have enough experience to advise you to whether her reproductive system feels normal or not.

Not settling on first service may occur for several reasons including: a fast ovulation cycle in which the sperm arrive too late, a low-grade uterine infection that wasn't detected, technician error, climate stress when extremely hot and humid days precede her estrus cycle, or simply because she didn't feel she wanted to be pregnant yet.

A second breeding is normal, but continued repeats of her estrus cycle could indicate an internal reproductive problem that needs to be checked by a veterinarian. If she has repeated heat cycles and it becomes five or six months since calving, you should have your veterinarian examine her to determine the problem.

A veterinarian is also qualified to perform a pregnancy check. Most can do an excellent job of determining whether your cow is pregnant forty days after breeding. This inexpensive service will allow you to continue through her lactation with the expectation of a calf at a determined date.

A rare situation that can occur is an abortion after a positive pregnancy check. The reasons vary for bovine abortion, but most occur because of fetal death. It could be the result of a severe illness or injury, a uterine torsion where it shuts off blood flow to the fetus, or other factors.

A loss of a fetus or a not fully developed calf will likely produce a slight increase in milk production for several days or weeks because her body may be tricked into thinking she has just delivered a newborn calf when the reality is much different. This will only be a temporary increase.

The likelihood of your cow aborting is very low. If abortion does occur, it is still usually possible to produce a later pregnancy. Consult your veterinarian immediately to assess the options available. If the cow is healthy, you may still wish to keep and work with her. This will dramatically revise your calving schedule, however. With the extended lactation, you may want to find a replacement for her or add another cow to your farm. It need not be the end of your milking experience.

Artificial insemination requires frozen bull semen. The semen is kept in a special insulated container with a vacuum-sealed lining. AI companies that sell semen can also help you maintain this unit. Always keep children away from the tank and avoid spills.

AI service requires a stainless-steel breeding tube into which the semen straw is inserted. The tip of the straw is snipped off. Then a plastic sheath is slipped over to hold the straw in place. The tube is then inserted through the vagina and cervix of the cow and the semen deposited in the uterus.

Natural service for cows is done with a bull. Although it may be more convenient than artificial insemination, there are inherent risks with keeping a bull on your farm. You need to weigh the benefits and risks of having a bull with the safety of you and your family.

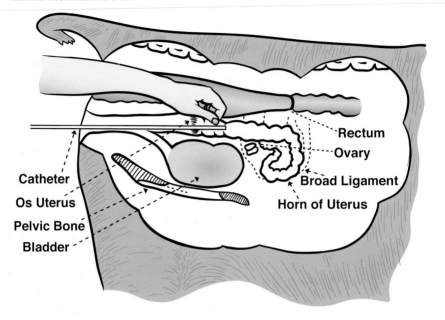

Rectum
Ovary
Broad Ligament
Horn of Uterus
Catheter
Os Uterus
Pelvic Bone
Bladder

Dry Period

Cows benefit from a rest from milking prior to calving. This is called the dry period during which she produces no milk. This rest period should begin about sixty days before the anticipated calving date. By drying up her milk production, her body can recover from the demands of her lactation. It gives her time to replenish many body reserves that she expended to produce milk in the previous lactation.

As her pregnancy progresses, your cow will begin to transfer many nutrients from milk production to help her calf grow. This is one reason for a drop in production during the last quarter of her lactation. Some cows cooperate easily while others may not want to stop milking. This has much to do with genetic potential, but you can design a safe program to help her quit on your own terms.

There are two basic actions required to get her milk flow to stop: quit milking her and drastically cut her feed and water intake. Regular, every day milking stimulates further production. In contrast, when her udder fills with milk from not being taken out, the pressure signals her body to stop producing milk.

This process should not require a long drying off. You can stop milking one day and wait two or three days before removing the milk she produced in between. Then extend the next complete milk out to four or five days, which should then be sufficient to complete the drying off. She may produce a minute amount after this, but it will be reabsorbed into her body over the next two months.

Withholding feed, and especially any grain, helps to decrease the nutrition available for making milk. It won't hurt her fetus because she will transfer whatever nutrition is available to maintaining the life of her unborn calf.

You can cut feed back easier than water. Depending on weather conditions, be sure she has access to water, especially if it is hot and humid, even though you are attempting to dry her off. Hot weather will help do it for you in this case. Water has no nutritional value, but it is needed to keep her cool. In cold weather, you may benefit from reducing her water supply when drying her off.

When you perform the last milk out, you will have two options: to infuse all quarters with an antibiotic or not to infuse them. Many dairy producers inject a tube of long-lasting antibiotic suspended in a gel-like substance in each teat after the last milking. Following this injection there is no more milk out since it would remove the intended protection from the antibiotic.

Family cows can be successfully dried off and kept through the dry period without the aid of antibiotics. Antibiotic residues may remain in her udder for a lengthy period, even as she approaches calving. Be aware that anyone in your household with reactions to antibiotics may be susceptible to them after she calves and you begin to use her milk again. Antibiotics should not be an issue once any remnants have been flushed from her system after the colostrum stage is over.

The calving that follows the dry period will bring you back full circle to where you started at the beginning of the previous lactation. It is the cycle of life.

Preparation for calving begins during a cow's dry period. Dry her up at least sixty days before the expected arrival of her calf. This dry period allows her to rest. And you will rest, too, because you won't have to milk her for those months.

Making Dairy Products
Say Cheese

■ ■ ■

NOW FOR THE FUN part. You've worked to feed your cow and learned how to milk her. You at least have enough to drink. Maybe you have more than you can use. How about making some cheese or butter, ice cream, or yogurt? It doesn't take that much extra milk to create these delicious dairy products. You'll be making the most of your extra milk.

Strain the Milk First

Raw milk can be consumed in its most natural state—straight from the cow. It's been done for thousands of years before refrigeration so it's nothing new. After finishing with the milking, you'll want to remove the straw, cow hair, dirt, or flies that found their way into your pail. It's an easy step to strain the milk.

You can use clean bed sheets, soft cotton cloth, cheesecloth, or muslin cut into squares large enough to line a colander. Farm supply stores usually sell small stainless-steel strainers, or you may be able to buy one at a local farm auction. These strainers use thin, wafer-like fiber pads held in place by a metal ring. The milk is poured into the strainer and then passes through the pad, filtering out any dirt or flies. These pads are usually discarded after each use, but if you are using the homemade cloths, you can wash and use them again.

Raw, unpasteurized milk is safe to drink if it comes from a cow that has tested negative for tuberculosis. Health issues with raw milk often center on unchecked bacterial growth that may occur during its handling coupled with human resistance, or lack thereof, to any bacteria present in the milk.

Proponents of raw milk consumption believe many healthful benefits are lost during the heating process of pasteurization. If you and your cow are healthy, there should be little reason to worry about consuming raw milk. Like any other food product, there are risks in consuming something in a raw state. You will need to weigh the benefits and risks from your own perspective, but understand that many farm youngsters have been raised for hundreds of years on raw milk.

Left: The fun part of home dairying begins with using the milk from your cow. You can produce tasty products for your family table.

Pasteurization an Option

Pasteurization of milk will eliminate concerns about harmful pathogens because the heating process kills most bacteria that are present. It does not, however, kill all the microscopic organisms that may be present, only those that may infect humans and most of those that cause spoilage.

There are three methods that can be used to pasteurize milk in your home: heating to a rolling boil for one second; heating to 170°F (77° C) for 15 seconds; or heating to 150°F (66° C) for 30 minutes. The last method is the best, as it avoids cooking flavors.

A simple home-pasteurizing unit can include a standard double-boiler or you can use two pots, one large and one smaller that fits inside the larger one. You will avoid scalding the milk on the bottom if you use this method rather than cooking the milk in a single pot. Place enough water in the large pot so that it will not boil dry in 30 minutes. Place the milk in the top pot. Heat the water to boiling and maintain a milk temperature of 150°F (66°C) by monitoring with a thermometer. Stir the milk to ensure an even temperature. After 30 minutes, remove the top pan containing the milk and set it aside in a sink of very cold water to cool it quickly. (You can use the same method for heating milk to 170°F (77°C) for 15 seconds or longer with similar results.) When the milk has reached room temperature, pour it into clean, sanitized jars or containers to store in the refrigerator.

Above: You can drink your cow's milk raw, or you can pasteurize it before drinking. Pasteurization heats the milk to destroy microorganisms that may cause disease or spoilage.

Left: Strain your cow's milk before using it. A metal strainer is placed over a jar, can, or tub. The milk is filtered through a thin fiber pad or cheesecloth to remove dirt, manure, hair, or bedding materials.

Making Dairy Products

There are many dairy products you can make with your cow's milk. The methods used to explain their production have variations. The following examples provide the basics. You are encouraged to study others to learn more.

Cream

Raw milk will separate overnight during refrigeration, and the cream will rise to the top of the container. You can skim off the cream and use it in your morning coffee or tea, or you can whip it for the tops of desserts or cereal.

Cream contains about 22 percent butterfat, leaving the rest of the milk with about 1 percent butterfat, hence the name 1 percent milk in the grocery store. Mechanical or electric separators are also available and can be used with warm milk to separate the cream instead of waiting for the natural separation to occur in the refrigerator. These machines require thorough cleaning and washing after each use to keep them sanitary.

With raw milk, cream will rise to the top of the jar overnight. This can be ladled off and used for coffee cream, over cereal, or whipped for dessert toppings.

Butter

Butter is easy to make. It requires only cream, a bowl for mixing, and a large jar with a lid. You can use a food processor or an electric mixer to take some of the arm work out of it.

Take the cream you've skimmed off the milk, and if it is chilled, let it reach a room temperature of 65 to 68°F (18 to 20°C). Cold cream will not churn very well. You need to let it warm up.

Use the food processor or electric mixer as if you were making whipped cream. Use plastic blades if they are available, instead of metal. As it stirs, the mix will go through a stage of forming firm peaks before becoming stiff. Reduce the speed and watch for a quick transformation into butter.

The butter will begin to clump together, and you can then drain off the remaining buttermilk, which can be used for baking or cooking or it can be fed to your family pets if no one wants to drink it. After the initial draining, use a large spoon to fold and press the butter to help remove any residual buttermilk. This will help it to stiffen and prevent the butter from becoming rancid.

Add clean, cold water to rinse the butter, and operate the blender on low speed for about one minute. Don't use warm or hot water because this may melt the butter, which will then run off with the water. Repeat this washing process until the water is clear. To remove the water, use a wooden spoon or paddle to press and shape the butter in the bowl. You can also use your fingers.

Salt the butter if you plan to use it soon. The amount used will depend on your tastes. About ½ teaspoon for each ½ pound of butter should satisfy most tastes. Be sure to thoroughly mix in the salt so that it doesn't create small salty pockets in the butter. You can also add garlic, herbs, or other seasonings if you want to make flavored butter.

Don't salt the butter if you plan to freeze it. The freezing process enhances the saltiness or flavor, and you may find it tasting different after freezing.

Butter-making begins with an electric butter churn.

Add warm cream to the churn and agitate. It will slowly transform into the first stage of butter. Pour into a cheesecloth or muslin. Drain any buttermilk that remains.

Remove cloth and butter and place in sink or bowl. Wash it with cool water until the water is clear.

Work the butter with your fingers to remove any trapped buttermilk.

After it's thawed out, you can add the salt and seasonings you want. Homemade butter can be stored for at least three months or longer in a freezer.

After mixing in the salt or flavors, shape it or press it into molds. Remove when set, and wrap in wax paper or put it into a dish to refrigerate.

Butter made from grass-fed cows will be a rich, deep yellow without food coloring added. It's a treat for your hard work.

Left: If salting, mix thoroughly to avoid salt pockets. When finished, roll the butter into balls or spoon into molds or shallow pans.

Below: Your finished product is ready to use.

Kids Can Make Butter Too!

Step 1. Fill small jar with cream and seal tightly.

Step 2. Shake, shake, and shake some more.

Step 3. Stop shaking when butter begins to form in jar.

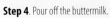

Step 4. Pour off the buttermilk.

Step 5. Rinse with cold water.

Step 6. Butter delight!

Yogurt making begins with heating the milk, then cooling it rapidly. Add the yogurt start culture and stir thoroughly.

After you have mixed in the culture, the milk will begin to thicken. Then pour it into warm glass jars or other heat-resistant containers.

To incubate the yogurt, set the jars on a warming plate or inside a dehydrator to keep warm. Let them set for about four hours and refrigerate to cool. This inhibits further development of the acids.

Yogurt

Yogurt is a fermented milk product. It requires the addition of a starter of active yogurt mixed culture bacteria. These bacteria produce lactic acid during the fermentation of lactose. They lower the pH, give yogurt its tart taste, and cause the milk protein to thicken. It also acts as a preservative because pathogenic bacteria cannot grow in acid conditions.

Yogurt is one of the fastest growing dairy product categories in the United States. It is easy to make at home. Adding fruit or berries to finished yogurt provides a delicious and nutritional treat. Yogurt can be made from whole, low-fat, or skim milk.

Pasteurize your home-produced milk before using it to make yogurt. The bacteria in raw milk would counteract the starter you put in and you wouldn't get a satisfactory product. After you have the milk pasteurized, you are ready to begin.

You may already have most of the equipment you need in your kitchen. Set these items out before you begin to keep the process moving smoothly. This includes a double-boiler that holds a minimum of five cups of milk; a liquid thermometer with a range between 100 to 300°F (37 to 148°C); a container of glass, crockery, or stainless steel that holds at least five cups; a large spoon and a bowl in which to stir the ingredients; and a commercial yogurt-maker with an electrically heated base that can maintain a constant temperature of 108 to 112°F (42 to 44°C) for incubation.

You will also need a starter. This initiates the bacterial process. An unflavored cultured yogurt starter is often available at health food stores. Less expensive commercial cultured yogurt can be found at many supermarkets. After you start making yogurt at home, you can save some of it for your next batch. But for best results, you will need to replenish a homemade culture every four or five batches with a commercial culture.

Pour one quart of cold, pasteurized milk into the top of the double-boiler, and stir in ⅓ cup of non-fat dry milk powder. Heat the milk to 200°F (93°C), but do not boil. Stir gently and hold for 10 minutes. Remove the top pan containing the milk, and place it in cold water to cool rapidly to 112 to 115°F (44 to 46°C). The temperature will fall rapidly. When you reach that temperature, remove the pan from the water.

Blend one cup of warm water with the yogurt starter culture in a bowl, and add it to the warm milk. The yogurt bacteria will thicken and flavor the mixture. The temperature should be between 110 to 112°F (43 to 44°C) when you pour it into a clean hot container. Then cover and place it in prepared incubator and close it.

Incubate for four hours. At the end of this time, the yogurt should be set. A longer incubation time will yield a more acidic flavor. Immediately refrigerate as the rapid cooling stops the development of acid.

Yogurt will keep for about ten days at a normal refrigeration temperature of 40°F (4°C) or lower. You can add berries, fruit, or nuts to give yourself a flavorful treat.

Remove from the incubator and stir in desired fruit, such as blackberries, raspberries, or cherries. It should keep in the refrigerator seven to ten days.

Other Options

There are two other methods that can simulate incubators if you prefer not buying a commercial yogurt-maker.

The first one involves lining an ice chest with aluminum foil and placing jars filled with 140°F (60°C) water inside. Then place the containers with the yogurt mixture in alongside and close the lid tightly. Keep a space between the jars of water and the containers of yogurt.

In the second method, pre-warm your oven to 200°F (93°C) and then turn it off. Use an oven thermometer to monitor the temperature and do not let it drop below 100°F (37°C). To maintain a temperature of 110 to 112°F (43 to 44°C), you will need to turn the oven on for short periods.

With either of these methods, you will need to keep an incubation time of four hours to create a good product.

Ice cream is a delicious and easy-to-make dairy treat with your cow's milk. *Shutterstock*

Ice Cream

Tasty ice cream. What could be better on a hot day, especially when you've made it with milk from your own cow?

You don't need expensive equipment to make homemade ice cream. You can use a hand-cranked or electric ice cream maker, or you can do it by hand. Any of these methods will allow you to make a smooth and creamy dessert.

There are two basic types of homemade ice cream: one that uses eggs for a custard style that is creamier, and one that doesn't contain any eggs. The eggs let the fats and water mix together better, which adds richness to the ice cream. They also help it hold up better against melting.

Ingredients for non-custard style ice cream:

1 c. milk
1 c. sugar
¼ tsp. salt
2 c. cream
½ tbsp vanilla extract

For custard style ice cream:

Add 5 eggs to non-custard mixture
Beat eggs, then mix with other ingredients until sugar dissolves

Pour milk into a heavy saucepan over medium heat. Bring milk to a gentle simmer at about 175°F (79°C), or until it bubbles around the edges. Remove from the heat. Add the sugar and salt. Stir until salt and sugar are completely dissolved. Add cream and vanilla and stir until well-blended.

Pour into a bowl and allow mixture to cool to room temperature or place bowl in ice-water bath for faster cooling. Once it is cooled, cover and allow mixture to age in refrigerator for a minimum of four hours or up to twenty-four hours. Aging will give it better whipping qualities, creating more body and a smoother texture.

After aging, remove from refrigerator and stir. The ice cream is now ready for the freezing process. Transfer the mixture to a bowl or container appropriate for freezing, cover tightly, and place in freezer for two hours.

Remove from freezer and beat with a hand blender, electric mixer, or a spoon to break up the crystals that are beginning to form. Cover and place back in freezer for two more hours.

Remove and beat again. It should thicken but be too soft to scoop. If it is not firm enough, return to freezer until it can be scooped.

Remove and pack the ice cream into a mold or container and place in freezer until solid enough to scoop and then serve. Yum.

Cheese-making

Your time has come. Not only to say cheese but to make it. Basic cheeses are relatively easy to make and range from soft to hard or young to aged. You can learn how to make many different kinds for your family or to sell at a farmers' market under your own label. Farm-produced cheeses are in demand. You can make a specialty or artisan cheese to sell. The opportunities are many, but it all begins with a basic knowledge of how to make a simple variety of cheeses. As your experience and expertise develop, you can work or experiment with more unusual varieties.

The following discussion outlines the basics of cheese-making and describes the processes to produce cheddar, a hard cheese, and cottage cheese, a soft cheese. There is much information available that delves deeply into many other varieties and specialty cheeses. You should study those that interest you. Many universities offer food science programs that include tutorials for cheese-making. A local cheese factory may provide an internship, or you may be able to make arrangements to work alongside established cheesemakers in a non-paying role to acquire a hands-on experience before attempting it on your own.

Most cheese in the United States is made from cow's milk. In recent years, however, there has been an explosion in the number of artisan cheesemakers who have branched out into using milk from goats, sheep, and, in some instances, water buffalo. Some cheesemakers mix goat or sheep milk with cow milk to produce desirable and very marketable cheese that combines the best characteristics of each.

To make cheese, the solid parts of the milk, called curd, must be separated from the liquid part, called whey. You will be using a bacteria culture called a starter, which is added to the warm milk. When thoroughly mixed, this will start the cheese-making process. To this mixture is added a rennet extract that acts as a curdling agent. This is mixed for three minutes before the milk is allowed to "set," during which time it begins to coagulate into a soft, semi-solid mass, which is the curd.

To more easily work with this mass of curd, it is cut into cubes. By this time, most of the whey has separated out of the curd. Both the liquid and the curd are cooked for about 20 minutes to remove any remaining whey inside it and to develop the proper degree of firmness and acidity. When the cooking ends, the whey is drained off and you will be left with a white mass of soft curd that will easily bind back into itself. From this point on, the cheese will quickly develop its character.

Cheesemakers vary this process or the ingredients to produce the different kinds of cheeses. They may vary the temperature, the amount of salt added, the amount of moisture drained from the curd, or the length of curing time. The kind of bacteria culture added will also determine different flavors.

Handcrafting your own artisan cheeses is one of the joys of keeping a cow. *Shutterstock*

Different cheeses require varying lengths of time to cure. Cheddar can be aged from sixty days to twenty years, or be eaten fresh soon after it's made. The shortest curing periods will produce cheese of mild flavor and smooth texture. Longer curing periods produce cheeses of sharper flavors and drier, more crumbly textures.

Don't think that you will be making twenty pounds of cheese from twenty pounds of milk. Milk of average butterfat (3.5 percent) and protein (3.2 percent) will generally yield about 10 percent, or in this case, twenty pounds of milk would yield two pounds of cheese. Some dairy breeds, such as the Jersey, are known for high average butterfat and protein percentages, and their milk may yield up to 11 or 12 percent.

The percentage yield may also be affected by your cow's diet, the climatic conditions she's exposed to, and, to a lesser extent, her stage of lactation; late lactation milk is typically higher in butterfat and protein percentages.

Say your cow produces five gallons of milk per day. This equals about forty pounds of milk, which will yield about four pounds of cheese. The rest of the thirty-six pounds is liquid or whey, which can be used in a variety of ways, such as whey cream or butter, because it is high in globular proteins and amino acids. It can also be fed to pigs, chickens, or pets.

The four pounds of cheese will be one day's production activity. The next day you will get another forty pounds of milk from your cow and so on. The realization that you have a large volume of milk that keeps coming may overwhelm you at first until you develop a plan to use the extra you don't need for daily drinking, cooking, or making dairy products. Fresh milk can be stored up to three days if properly cooled. Longer than that and the putrefactive bacteria that produce rancidity will slowly increase and ruin the entire batch.

If you have proper cooling equipment that maintains a constant temperature at 42°F (5°C), you can store your milk for several days and then make a larger batch of cheese. This may alter your schedule from everyday cheese production to two or three times a week. During this cool storage time, however, you will need to stir the milk to ensure an even flavoring and cooling.

Milk that is used for cheese-making should be pasteurized if you plan to eat the cheese before ninety days. Raw, unpasteurized milk can make excellent cheese, but it then must be aged for a minimum of ninety days before it can be legally sold. Your family can eat it anytime, however. The aging process destroys any harmful pathogens that survive the cooking process.

Set up your equipment before you begin to make hard cheese. Once the process starts you need to be prepared and stay close by.

Equipment Needed

- Heavy stainless steel pot with lid (NOT aluminum)
- Measuring cups from ¼ cup to 1 quart
- Thermometer with range of 32 to 225°F (0–100°C)
- Whisk for mixing starter and rennet
- Cheesecloth to catch curd and allow whey to drain
- Cheese press (required for hard cheese such as cheddar)
- Heat source such as stove
- Cutting knife
- Spoons
- Strainer

Ingredients Needed

5 gallons pasteurized milk*
3.6 grams 0.02 percent concentrate calcium chloride (roughly ¾ tsp.)
¼ c. water
Starter
Yellow food coloring, if desired
Rennet
4 tbsp. non-iodized salt
*Use any amount of milk but adjust the ingredient ratio accordingly.

After pasteurizing the milk, warm it to 68°F (20°C). Add the calcium chloride to ¼ cup water until dissolved. Then mix thoroughly with the milk. Calcium chloride needs to be added because pasteurization removes calcium from the milk and you need to replace it.

Once the milk is warmed, add the starter in recommended amount. This can be purchased commercially. There will be guidelines with it. You can use cultured buttermilk or yogurt as starters, but be careful not to add too much. Too much will begin to thicken the milk before the rennet is added. The starter culture acidifies the milk and turns the sugars (lactose) into lactic acid. This process is referred to as ripening the milk.

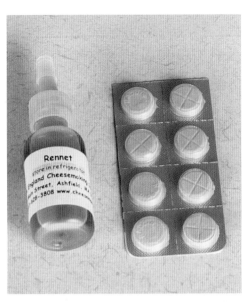

Rennet is a curdling agent for cheese-making. It used to be made from the stomach lining of a calf, but most rennets produced today have non-animal origins. They may be in liquid or tablet form.

Stir and mix these ingredients. You can use a hand mixer at a very low speed to help in mixing, being sure to thoroughly stir the bottom and sides of the pot. If you wish to add yellow food coloring to give the cheese a typical yellow color, it should be done at this point.

When the milk and other ingredients are completely mixed, add the rennet. This is commonly available in supermarkets in the pudding section or can be ordered from cheesemaker supply houses. Either tablets or liquid rennet can be used. Use the recommended amount listed on the package. One tablet equals about one teaspoon of liquid rennet and coagulates five gallons of inoculated milk. Mix or stir thoroughly for about three minutes, and then let set so it will coagulate for about thirty minutes.

To test the firmness of the curd, dip your clean forefinger into it at an angle and then lift. If the curd is stiff and breaks cleanly over your finger, it is ready to cut. If the milk is still gelatinous and flows across your finger, let it sit until a clean break is obtained. Do not stir it!

Once a clean break occurs, the curd is ready for cutting. Use a long knife and start at the edge and cut straight down to the bottom of the pot. Cut repeatedly parallel to your first cut, and after cutting both ways vertically, use the knife at an angle following the previous slice marks in both directions to cut it into cubes from top to bottom. Cutting the curd into cubes helps in separating it from the whey because there is more surface area for the whey to escape.

To make cheese, start by heating the milk in a double boiler. Add calcium chloride, a starter, and rennet at the appropriate times. Thoroughly mix or stir for three minutes. Then let set undisturbed for twenty to thirty minutes. The soft curd mass will rise to the top.

When ready, cut the curd with a long knife in two directions. Try to make cubes when cutting so that more whey will be removed during the cooking process.

How Much Salt?

If you are unsure about the amount of salt to be used, the following is a guide:

40 lbs. milk × .10 yield = 4 lbs. cheese × .25 salt = 0.1 lbs. = 1.6 oz. = 48 g. = 4 tbsp

Place the pot over a low heat and stir the curd gently with a clean hand, wiping any curd residue off the sides and bottom of the pot. Cut any large curds as they appear, but do not mash or squeeze them.

Stir by hand for about fifteen minutes to prevent clumping. If you plan to make cottage cheese with some of this curd, now is the time to set aside the amount you want to use. Removing curd at this point won't affect the outcome of the cheese; it will only affect the total volume of cheese you get at the end.

Raise the temperature to 102°F (39°C) and maintain for twenty minutes, stirring constantly with a large spoon. Make sure to completely stir the bottom. The curd should be firm but not hard by the time you begin to drain off the whey.

Raise the temperature to 102°F (38°C) and hold for twenty minutes. You should stir the curd and cut apart any large pieces. Don't use your fingers to crush them.

Cheese-making Cautions

There are three points of caution with cheese-making. First, do not use aluminum pots for heating in the cheese-making process because the acidifying milk can dissolve aluminum and contaminate the milk. Second, sterilize the pot before each batch or use. Finally, if the curd floats to the top and stay there while cooking, it could be a sign that you have a contaminant in your starter that produces gas or that your milk was contaminated. This is one reason why you should purchase fresh starter rather than use another product such as buttermilk or yogurt. Bacteria that form bubbles may be useful in cheeses, such as Swiss cheese, where captured gas creates the distinctive holes in it. Bubbles don't necessarily ruin your cheese, but they increase the chances of producing off flavors. Many CO_2 forming bacteria are non-pathogenic. To be safe, age any cheese exhibiting gas for at least sixty days because pathogens do not survive extended aging.

After the cooking is finished, turn off the heat and remove to let it cool. You can slowly drain off the whey by pouring it into cheesecloth over a container. If you want to keep the whey for later use, catch it in a sanitized container.

Below: Allow the cheese to drain for five minutes. During this time set up the press you will use later. The excess whey can be fed to pets or other animals.

Remove the cheese and cloth from the colander. Place in a large bowl.

Break the curd into crumbles with your fingers. This will allow any remaining whey trapped inside to drain. You can add salt or herbs as you mix the curd in the bowl.

Prepare your cheese press by lining it with cheesecloth. Set the press over a clean grate and pan to catch the whey as it is pressed out.

Transfer the curd from the bowl into the press. You can use your fingers to press it in as you fill it. You may need two presses if you make more than fits into one, or the excess can be eaten fresh.

Turn off the heat after twenty minutes, remove the pot from the heat source, and let it sit to cool. Line the strainer with cheesecloth and place over a clean pail or bucket if the whey is to be used for the household. Slowly pour the whey and curd into the strainer, allowing the whey to drain completely.

You can wrap the curd into a ball inside the cheesecloth and hang it while the remaining whey drains out by gravity. After five minutes, remove the warm curd and spread out the cloth on a table and crumble the curd with your fingers. Sprinkle salt over the curd (about 2.5 percent of the total weight), or add herbs if desired. Use food-grade salt that is not iodized. Mix the salt in with your hands. When it is thoroughly mixed and dissolved, pour off any accumulated whey.

If you use an electric cheese press to help remove the whey, apply it at the equivalent of about 18 pounds pressure. If using a homemade press, you will need to apply some pressure to squeeze any remaining whey out of the curd to promote aging. A sturdy, clean, round container such as a large coffee can that has the ends removed may work well. You need both ends open so the whey can escape the curd at one end as pressure is applied at the other. Lining the container with cheesecloth helps in removing the pressed cheese once the whey is pressed out.

Place the still warm curds in the cloth and cover the open ends of the container with the cloth and place a perforated lid on top. Turn it over and repeat this at the other end. Then apply a weight on the top lid and let gravity press it down. Let it stand for one to two hours at room temperature. If you plan to routinely make cheese, you may want to invest in a cheese press as it will help produce a more consistent product.

When time is up, remove the press, remove the cloth, and rub the outside of the cheese with salt if desired. Rewrap it in fresh cloth. Place the wrapped cheese on a rack in the refrigerator, but do not let it freeze. A dry, yellowish rind will form in one to two weeks and you can then dip it into melted wax used for food processing and then age it for the desired time.

Place the weight on top of the curd and allow gravity to slowly press out the whey. If you plan to routinely make cheese, you may want to invest in a good electric cheese press. This will help create a more uniform product.

Cottage Cheese

The process for making cottage cheese starts like that for making hard cheese. Assemble the milk, calcium chloride, water, starter, and rennet to create cheese curd as outlined above. You can use the same steps until you reach the stage where you have cut the curd into cubes and stirred over low heat for fifteen minutes. Now you are ready to use the curd to make cottage cheese. Raise the temperature to 102°F (39°C), and drain off the whey, but flush with cool water to keep the small curds from binding back together and to allow them to separate. Then add salt to taste, and presto, you've made cottage cheese!

Cheese-making can be a satisfying and delicious experience and one that involves the whole family. A little practice and soon you'll be able to develop artisan cheeses to enhance your meals.

1. Cottage cheese-making starts with taking one pint of milk from the evening milking and letting it stand overnight.

2. Add one pint of fresh milk from the morning milking to the standing milk and leave sit at room temperature for eight to twelve hours.

3. Break up curd with a whisk and then pour in two cups boiling water. Stir thoroughly with wooden spoon.

4. Test for firmness with fingers.

5. Strain through cheesecloth or strainer.

6. Wash with clean water until clear. Salt to taste, and it's ready to eat.

CHAPTER 10

Health
Taking Care of Business

■ ■ ■

A H E A L T H Y   C O W is a productive cow. While she may be quite self-sufficient, you will still need to provide some day-to-day care and attention for her safety and well-being. This includes daily observation, which is easy if you are milking your cow twice a day. A healthy, independent cow is the ideal situation that you hope for and many days will be like that. But there will be days that are not ideal and those days are critical to both of you. Being alert to any illness, lameness, or other physical or digestive ailments will help avoid many serious health problems.

Good cow care involves proper feeding and adequate nutrition. It includes close care at calving time and attention to things that make for a comfortable and humane existence, such as proper housing and shelter.

Keeping Your Cow Comfortable

Cows respond to comfort and discomfort and good treatment and maltreatment. The best and most alert dairy cow handlers have the perception and ability to read the body language in their cattle. By watching the movements of their cow, they can instinctively tell whether she is comfortable. If not, steps are taken to improve the situation. You can learn this too with time and experience.

Bellowing, kicking, and general agitation are signs that the situation needs your immediate attention, not six hours later. A cow shows signs of illness in several ways, including a drooping head or ears, a listless tail, or lying down and getting up numerous times, particularly if it is unrelated to calving. Cold ears, when touched, should alert you that she may be getting sick.

The normal physical processes of your cow require a body temperature maintained within narrow limits—between 100 to 103°F (37 to 39°C). The normal core body temperature of a healthy, resting cow is about 101.5°F (38.6°C).

Left: Healthy cows are productive cows. Being alert and curious with bright eyes is a sign of good health. You will become a cow expert the more you work with them.

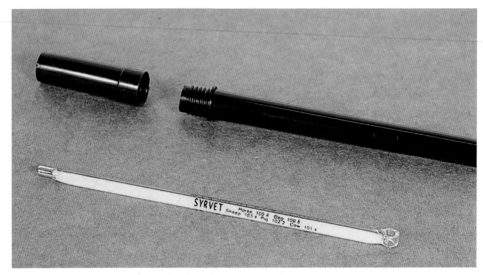

A rectal thermometer should be a part of your dairy cabinet. Keep it encased when not in use.

Although every cow's metabolism is different, this normal temperature does not vary much. Pay attention to anything outside this range. Keep a veterinary rectal thermometer available to do checks if needed.

Environment, weather, and climate have some effect on a cow's body temperature. Time of day and level of activity also have an effect. Morning temperatures are generally lower due to her coming off a time of resting and cooler nights. Her temperature may be slightly higher in late afternoon because of a day of activity.

Humidity, wind, and the sun along with the air temperature affect a cow's ability to maintain a steady temperature. As the air temperature rises on a hot summer day, her stress increases as her body works to expel the excess heat. Providing shade or shelter keeps her out of direct sunlight, which helps mitigate some of those conditions.

Humidity is likely a greater detriment to your cow than sun. Shade can counteract the sun's rays, but when humidity levels are high there are few ways to keep her cool. Setting up a water mist may help during the hottest part of the day; she is likely to lie down in the puddle of water that forms in an attempt to cool off. Keeping her in a pen and using a large fan can help. Lack of appetite usually results from high body temperatures, and this results in lower milk production.

In extreme cases, a cow's core body temperature can rise to dangerous levels. When this happens, immediate action must be taken to cool her down or heatstroke can occur. In times of extreme heat or humidity, make sure your cow has full access to clean water.

A cow has 360-degree vision, which means she is alert to all kinds of movement. Dark shadows where she cannot distinguish between certain objects can lead to a situation where she responds differently than in a well-lit area. This is particularly important to remember when she first arrives at your farm because everything is unfamiliar to her. Developing a familiar routine for her helps avoid surprises during evenings and winter months when there is less light. A cow is a creature of habit and her familiarity with an area allows her to respond in a way she feels comfortable.

Managing Her Health

Good feed, a clean stable, and comfortable surroundings all contribute to keeping your cow healthy and happy. Still, animals get sick even with the best of care. The key is to minimize the number and severity of those illnesses to the greatest extent possible. Daily observation is the first step in good health management.

If your cow gets sick and you feel the need to call in a veterinarian for assistance, do so. However, you can treat most cattle illnesses without any specialized education. Experience will be your best teacher. You can also gain insight and understanding by reading books on cattle health.

There may be a time when veterinary assistance is essential to treat an illness, such as a high fever. A licensed veterinarian is required to administer vaccination injections, medical euthanasia, and certain antibiotics or steroids.

There may be a critical time, however, when you need to use your own initiative, such as during a case of bloat where time for treating it is extremely short and a call to a veterinarian may be too late.

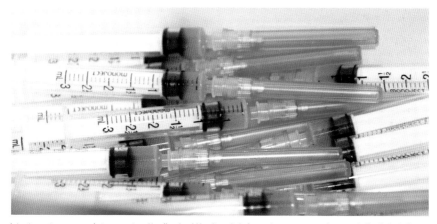

Injection syringes are made in many sizes. Needles should be discarded after each use to prevent the spread of body fluids from one animal to another. Clean the syringe after use and before storing.

Treatment Options

If your cow is sick, there are two basic treatment options: conventional, which generally involves the use of antibiotics to relieve the symptoms of the illness or disease, and alternative, which involves homeopathic or herbal treatments without the use of antibiotics or steroids. Like other dairy producers, you will want to use the one that works best for your situation and aligns with your management philosophy.

Antibiotics

Antibiotic use started in the mid-1940s, and by the early 1950s, the food-producing industry saw the introduction of it in commercial feed for cattle. For the past fifty years, antibiotic use has served three purposes in cattle: as a therapy to treat an individual illness, as a prophylaxis to prevent illness in advance, and as a performance enhancement to increase feed conversion or growth rate.

Antibiotics can be used to help reduce the suffering and distress in animals caused by bacterial diseases. By speeding the recovery of an infected animal, the economic loss can be checked significantly.

When antibiotics are used responsibly, they can be an essential element in the fight against animal diseases. When they are not, they can lead to long-term problems, such as bacterial resistance. One study has shown that every year, 25 million pounds of antibiotics, or roughly 70 percent of the total antibiotic production in the United States, are fed to chickens, pigs, and cattle for non-therapeutic purposes, such as growth stimulation. This report also showed that the quantities of antibiotics used in animal agriculture dwarf those used in human medicine. Non-therapeutic livestock use accounts for eight times more antibiotics than human medicine.

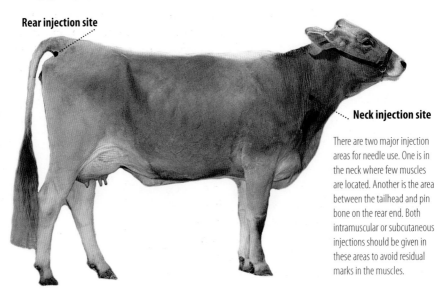

Rear injection site

Neck injection site

There are two major injection areas for needle use. One is in the neck where few muscles are located. Another is the area between the tailhead and pin bone on the rear end. Both intramuscular or subcutaneous injections should be given in these areas to avoid residual marks in the muscles.

Requires Immediate Help

- Heavy and uncontrolled breathing
- Cow down and on her side, unable to get up or roll over
- Cow has gone into labor but has given up; calf may or may not be partly visible
- Prolapsed uterus following calving
- Bloat
- Temperature of 104°F (40°C) and above; may be in conjunction with other signs
- Subnormal temperature or sudden drop in temperature; may be acute mastitis
- Calf with white, bloody, or watery diarrhea

Get Help Soon

- Cow up or down, immobile, but chewing her cud
- Cow has refused water or feed for six hours
- Bloody urine or feces
- Any behavior that interferes with her normal routine of drinking, eating, resting, and breathing
- Hard swelling in udder or change in appearance or consistency of milk
- Sudden, unexplained major drop in production

The resistance of bacteria to drugs, such as antibiotics, has been documented in several studies. The more antibiotics are used, the greater chances of an antibiotic residue entering the food chain that can affect the general population. New antibiotics are being manufactured to stay ahead of the resistant strains of bacteria that are evolving as well. Drug-resistant bacteria are now a major health concern—an irony because they fed on and evolved from the antibiotics that were meant to kill them.

This information doesn't mean that antibiotics shouldn't be used. It means that they should be used judiciously and only when warranted. In specific cases, such as Johne's disease, it is better for animals to be culled rather than put through any treatment program because in most Johne's cases, the treatment is ineffective and uneconomical.

Antibiotics are used for ailments such as fevers, mastitis, and cuts. They should be properly labeled and stored in a refrigerator at the prescribed temperature.

Mastitis treatments are generally enclosed in plastic tubes. Alcohol pads are part of the kit and are used to swab the teat end before the tube is inserted.

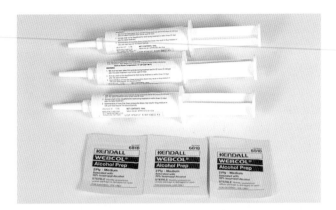

To infuse antibiotics into an udder, first thoroughly swab the teat end and opening with an alcohol pad. Remove the plastic covering on the tip of the tube and insert the tip through the end of the teat and into the teat canal. Press the plunger and push the antibiotic into the teat. Remove the tube and dip the teat end with iodine or dairy disinfectant. Slowly massage the infusion up into the udder.

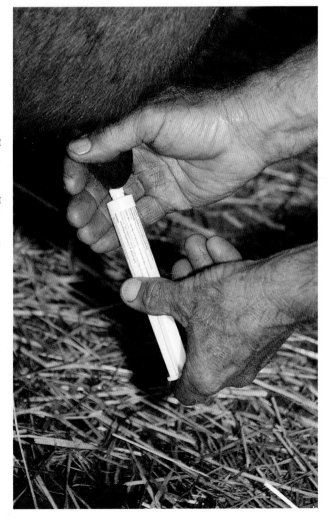

For any animal that has received antibiotic treatment, its milk (or meat) must be withheld from the market for a specific time. The withholding times must be followed in order to prevent antibiotic residues remaining in the muscles and possibly causing an allergic reaction by human consumption of meat products.

Alternative Treatments

Maybe your philosophy rejects the use of antibiotics. Alternative methods are available for you to use, including homeopathic and herbal treatments. These methods are ideal for those considering organic, sustainable, or biological farming methods with their animals and marketing products produced from them.

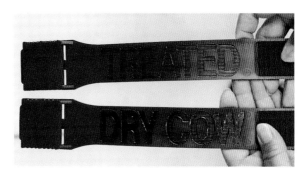

Tags are used to indicate any cow treated with antibiotics. Do not use antibiotic-treated milk for human consumption. Always follow the recommended withdrawal times on the label.

In years past, these homeopathic alternatives had been pushed aside for the quick fix of antibiotics that became inexpensive, readily available, and demonstrated a fast response in the affected animal. Why wouldn't you use them if they worked so well? In some cases, it was thought that using antibiotics routinely could replace good management practices.

Homeopathy is based in part on the idea that bacteria are not necessarily a bad thing and that they do not need to be destroyed. It is not the bacteria that needs to be treated but the animal's reaction to it.

Homeopathy treatment involves the natural stimulation of the animal's immune system so that it can fight off the bacteria that might otherwise cause a disease. Antibiotics have a suppressing effect on animal immune systems while they are fighting the bacteria. But they don't differentiate between fighting the good or bad bacteria because they fight both.

Letting an animal use its own body to fight disease-causing bacteria provides more benefits in the long run because its whole physiological system is in better condition. Building up the health and strength in the animal will result in good growth rates and milk production by itself. A healthy cow produces healthy milk and meat.

You can use two approaches when adapting a homeopathic system on your farm: preventative and therapeutic. Therapeutic is the emergency treatment of individual cases. Common sense tells you that a program of preventive medicine is better than treatment, and homeopathy is well-suited to this approach. For example, homeopathy can be used to support the development of a calf's immune system before it is born by working with the pregnant mother. The first six months of a calf's life are the most important and will determine, to a large extent, its health later on. A sickly calf does not become a healthy cow overnight but a healthy, vital calf has the ability to stay healthy.

Homeopathic products are administered by means of a remedy. These are derived from all-natural sources, including animal, mineral, or plant, and their preparation is made by a qualified homeopathic pharmacist. A remedy is used in doses that are generally marketed in one-gram vials and based on a system of potencies. In tincture form, the remedy is added to a sugar-granule base and allowed to soak, after which it becomes stable and can remain active for months. If properly stored, some remedies can remain active for several years. These vials usually contain sufficient materials for several doses, or administrations, for the animal. A dose may consist of five or six small pellets given at one time, depending on the condition being treated, and may have several applications during one day.

The remedy is placed directly on the animal's tongue and it is dissolved by saliva. The granules do not need to be swallowed because homeopathic remedies can be absorbed through the palate or tongue.

Homeopathic potencies are determined by the dilutions made from the refined crude product. This refinement develops the inherent properties of the remedy used for the specific needs of the animal.

Homeopathy is a legitimate route for treatment of animals, but it is not a substitute for good husbandry or preventive measures, such as good nutrition, air quality, and proper sanitation.

Herbal Treatments

Herbal treatments are the second alternative available for your use. Remedies are preparations made from a single plant or a range of plants. Different methods are used to apply herbal treatments, depending on the perceived cause of the disease condition. These applications can be made from infusions, powders, pastes, or juices from fresh plant material. You can apply topical treatments for skin conditions while powders can be rubbed into cuts or incisions. Oral drenches can be used to treat systemic problems, and drops can treat eyes and ears. There is much information available about homeopathic and herbal treatment protocols from alternative health stores or books published on these topics.

Common Health Problems

Before bringing your family cow home, you should find a local veterinarian who includes cattle in his or her practice. Develop an acquaintance before assistance is required. You will be able to explain your farming situation and you may be offered suggestions even before you bring your cow onto your farm.

You may prefer to try homeopathic remedies for prevention or treatment of bovine ailments before rushing to use antibiotics. *Shutterstock*

Releasing a hungry cow into lush spring fields can lead to bloat, a serious disorder that can cause death if untreated. Ward off bloat by feeding your cow dry hay before sending her to pasture. *Shutterstock*

Pastured cows appear to require less veterinary care and exhibit fewer illnesses. They live in the open air unencumbered by the effects of unnatural housing. They move about more than confined animals.

You can handle some common health problems on your own without needing veterinary assistance. These include bloat, scours or diarrhea, pneumonia, clostridial diseases, parasites, and skin infections. Young calves and heifers are often more susceptible to several of these than older cows because of their immature immune systems.

Bloat

Bloat can kill your cow quickly. It is not a disease; it is a consequence of fermentation. Bloat occurs when the gasses in the rumen are unable to escape. Cows normally spend about eight hours a day chewing their cud and belching. If the belching process that lets gasses escape is severely hindered, it can create a very serious condition: bloat. It requires immediate attention.

Bloat can occur at almost any time of year in any kind of pasture, although it most often occurs on pure stands of young, lush alfalfa hay. Young plants are often low in fiber content, and it is fiber that stimulates the rumen to develop normal fermentation. When low-fiber materials ferment, they create a slime-froth, which forms at the top of the rumen contents. This prevents the normal fermentation gasses from being released by belching.

As the gas pressure increases, the cow stops chewing and the rumen begins to swell because she's not belching. As the gas slowly builds, all the internal organs—including the heart, lungs, liver, and intestines—are being squeezed by the enlarged and tightly pressured rumen.

If bloat is left untreated, your cow can die. In some cases, death is due to a heart attack because the pressure against the heart doesn't allow it to beat regularly. But in most cases, the cow dies from asphyxiation because the pressure inside the body cavity has squeezed all the air from the lungs.

Bloat can be treated with an oral drench of vegetable oil. The oil disperses the gas bubbles and allows the normal return of belching. Sometimes a small rope slipped across the cow's tongue, similar to a horse bit, and tied at the top of her head will force her to chew the rope. Both oil and a rope can be used at the same time to relieve the gas effects. After several minutes, the gasses will escape in the same manner as if she were chewing her cud.

In the most extreme case, the cow's upper left side may be bulging from extreme pressure. You may need to slice through the hide with a knife or puncture it with any sharp object to eliminate the gas and pressure rapidly. You can take steps later with a veterinarian's assistance to sew the hole shut and ensure that fermenting rumen materials didn't infect the body cavity. But at least you will have saved your cow.

You can minimize the chances of your cow developing a bloat condition by keeping her out of the pasture if she is extremely hungry. This can occur if she hasn't had anything to eat for several hours and goes to the pasture and gorges herself on grass. Overeating in a short period is one of the main causes of bloat. Also be aware of early morning frost on plants because ingesting too much frosty plant material can cause the formation of gas bubbles. You can decrease a cow's chances of developing bloat by feeding her dry hay before she returns to pasture. Dry fiber retards a rapid formation of gas in the rumen because it acts like a sponge until digested.

Pneumonia

Pneumonia is most common in weaned calves that have been kept in wet bedding or areas with cold drafts. But cows can experience pneumonia too. It can occur due to stress, changes in weather, and infectious agents all occurring at similar times. Develop a vaccination program with your veterinarian that includes IBR (infectious bovine rhinotracheitis), PI3 (parainfluenza type 3), BRSV (bovine respiratory syncytial virus), and BVD (bovine virus diarrhea) to provide a broad-spectrum defense against respiratory diseases in calves and cow. You may also include a five-way Leptospira, which will prevent waterborne infections related to organ dysfunction. The best preventative of pneumonia, however, is a dry, bedded, draft-free stable that has good air circulation.

Clostridial Diseases

Clostridial diseases are a group of related infections that may cause sudden death, especially in young cattle. These diseases include blackleg, enterotoxemia, malignant edema, and black death. They can be prevented with vaccines that stimulate the immune system. If your veterinarian identifies any of these as problems in your region, you should vaccinate your cattle for them. They can receive boosters at a later date.

Scours

A common ailment of newborn calves or young calves is diarrhea, commonly known as scours in livestock. If diarrhea occurs in an older cow, it is generally a symptom of another, more pressing problem. In calves, scours is a problem in and of itself. Scours causes dehydration, which is a prime cause of death in calves.

The prevention of scours is easier and more successful than treatment. The best preventive measure in calves is to provide pregnant cows with adequate nutrition, especially in the final two months before calving. It is important that the calf receives colostrum milk in the first six hours of its life. The antibodies in colostrum help develop protection against scours. Colostrum can be absorbed quickly through a calf's stomach and intestine during the first six to eight hours

A calf's best protection against potentially deadly scours is to receive the mother's colostrum within the first six hours of birth.

of life. After the first twenty-four hours pass, the absorption rate of the antibodies decreases dramatically. It's a short but very important window for the calf.

If your cow or her calf develop diarrhea, you can use electrolytes and fluid given orally or intravenously each hour to provide the most effective treatment.

Parasites

Internal and external parasites pose a problem for cattle. Cattle grazing on pasture tend to have a greater problem with internal parasites than cows that are not grazed. Parasite eggs are often attached to the leaves of grasses and hays, and when the cow grazes these plants, she ingests the eggs. Silages tend not to harbor parasite eggs because the heating process destroys them. Deworming may be needed to minimize the parasite load in the cow's body if it becomes severe. External parasites, including lice and horn flies, can be controlled with powders and other readily available products.

Fly control is important. There are many products available. They are packaged as powders, sprays, or aerosols. Be careful. These products are poisons and need to be used in areas with good ventilation.

Brucellosis

Brucellosis, sometimes called Bang's disease, is a bacterial infection that can cause spontaneous abortion in bovine. It mainly affects cattle and the American buffalo. Infected cows exhibit symptoms that may include abortion during the last trimester of pregnancy, retained afterbirth, or weak calves at birth. Infected cows usually abort only once in their life because they develop a level of immunity to further infections.

Brucellosis is a reportable disease to the state agriculture department by local veterinarians who diagnose it. A single suspected case will cause your herd to be immediately quarantined if you have more than one animal. It is a highly contagious disease and if the subsequent brucellosis test is positive, she will be sent to slaughter. Her meat can still enter the food chain because this is a reproductive disease and not a systemic one.

Bang's disease can cause a disease in humans called undulant fever, which is transmitted by drinking unpasteurized (raw) milk from an infected animal.

The best prevention is buying an animal that has been vaccinated against brucellosis. You can identify these animals because they will have a state-issued, colored ear tag that is made of metal and numbered. Some animals may lose their vaccination tags, but there is a secondary requirement to vaccination that you can look for: an ear tattoo. This is located inside either ear and will have several numbers and letters as a code identifying the date the animal was vaccinated. If you see an ear tattoo but no tag, you can be fairly confident that she has been vaccinated. In older cows, this tattoo should look old, not of recent vintage. If anything with the tattoo or ear tag appears abnormal, consider another cow.

Ear tags and ear tattoos indicate that a cow has been properly vaccinated against the contagious disease brucellosis.

Many dairy producers routinely vaccinate their calves against brucellosis, and you should too. This gives you the assurance that you won't have this problem with your cow.

Tuberculosis

Tuberculosis (TB) is a contagious disease that can affect cattle and be passed to humans. It is caused by three specific types of bacteria that are part of the mycobacterium group. Bovine tuberculosis has the greatest host range and can infect all warm-blooded mammals. Infected animals may appear normal most of the time but cough a lot. They often become listless and emaciated with low-grade fluctuating fevers during the late stages of the disease. Because the visual signs aren't always reliable, TB is usually diagnosed on a farm with a tuberculin skin test.

TB can be transmitted from a cow to humans by drinking unpasteurized (raw) milk from an infected animal. TB is a reportable disease, and if you or your veterinarian suspects your cow has an infection, a tuberculin test should be made. If the test is positive, she is sent to slaughter and other cattle on the farm are then tested.

The best precaution is to buy a cow from a reputable dealer or farmer or from a state that has a TB-free status.

Pinkeye

A highly contagious infection is pinkeye (infectious bovine keratoconjunctivitis) that affects the eyes of cattle. Its name describes the reddening of the eye caused by bacteria. If left untreated, it can render the animal blind in the affected eye. This condition usually occurs in warm months and appears to be transmitted from one animal to another by flies that congregate around their eyes. Pinkeye can be treated with powders, ointments, and antibiotics, or with the use of homeopathic remedies. You can get pinkeye from an infected animal. Be sure to use good hygiene during and after treating any case of pinkeye to avoid contaminating yourself.

Ringworm

Ringworm is caused by a fungus, not a worm, contrary to what its name may suggest. It is a contagious infection of the skin. It is easily recognized by the circular white-encrusted spots on the skin. It is difficult to treat. Fungus infections are harder to kill off than bacterial infections. Treatments include disinfectants, such as iodine and glycerin, that penetrate the scabs and saturate the spots. The

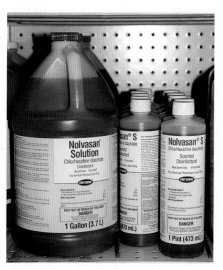

best defense is to avoid bringing infected animals onto your farm. Maintaining clean surroundings, pens, and having a well-ventilated barn are the best ways of controlling it. One of the oddities of ringworm is that the fungus causing it can lie dormant for several years and seem to be gone until conditions become right for an outbreak.

Ringworm can be easily transferred from animal to human. You need to use care around an infected animal or during treatment of it. Using disposable latex gloves when treating an infection will help prevent spreading it to you.

Many topical treatments are available for surface infections, such as ringworm and warts. These include iodine-based solutions like the one shown above.

Warts

There are four different types of papilloma virus that cause warts. They usually affect young calves and heifers under two years of age. They are often more of an appearance problem than a physical one. Vaccines are available, but most warts will disappear on their own if left alone. Their sudden appearance usually coincides with a sudden disappearance.

Metabolic and Environmental Disorders

Two different groups of disorders have the potential to affect your cow. Metabolic disorders include milk fever (see page 137), ketosis, fatty liver syndrome, and grass tetany. Environmental disorders can include poisoning from chemicals or plants, toxins caused by moldy feed, acidosis, hardware disease, or a displaced abomasum.

This cow has healthy eyes. If your cow contracts pinkeye, treat it immediately and be sure not to transmit the infection to yourself.

Ketosis (acetonemia) occurs when a cow's liver has been depleted of stored glycogen. This condition is most frequently observed in well-conditioned cows two to six weeks after calving. Energy from fat is mobilized to the liver and produces ketone bodies that are burned throughout her body. Fat reserves that are used too rapidly with insufficient carbohydrates available cause the cow to drop in milk production in only a few days. This can be treated with propylene glycol or sodium propionate used as an oral drench. You can test for it in her milk or urine with a ketosis kit.

Fatty liver syndrome is the accumulation of fat in the liver. Cows that are overweight at calving are most susceptible. Dairy cows do not normally store fat in their livers. This condition occurs when a cow breaks down more fat than her liver can properly process. This leads to reduced appetite and milk yield. In severe cases, it can cause her death. When the liver is not working at its normal level, other body functions are affected. Secondary conditions may occur, including ketosis and a displaced abomasum. Prevention is more important because the treatment process can be long and is often ineffective. Treatment can include glucose, propylene glycol, and corticosteroids. Feeding a balanced dry cow ration and preventing excessive weight gain during the dry period will help prevent fatty liver syndrome.

Grass tetany (hypomagnesium) is usually observed in cows that graze on lush grass pastures high in nitrogen, which results in low absorption of magnesium. It happens most often when grazing very young grass pastures since these have a

Ketosis and milk fever can be successfully treated if caught in time. Propylene glycol or sodium propionate is used as an oral drench for treating ketosis. Calcium is the best treatment for milk fever and is given intravenously. These products are available from farm supply stores or your local veterinarian. *Marcus Hasheider*

lower magnesium level than older grasses. The signs of a cow having grass tetany include walking with a stiff gait, falling, and going into convulsions. Death can occur. Prevention is a better option than treatment, which is usually done with an injection of calcium and magnesium intravenously followed by an injection under the skin. The best prevention is to watch your cow when pasturing her on grass fields that may have been heavily fertilized with nitrogen. Heavy nitrogen use on young pastures reduces the levels and amount of magnesium available to the plants. Adding two ounces of magnesium oxide daily to her feed during the young grass period should help to avoid this condition. This treatment may have to last for two to three weeks. As the grass matures, you can cease supplementing magnesium oxide.

Poisonings are environmental disorders that can have serious consequences for your cow. Poisoning can be by chemicals or plants. Chemical poisoning should not be a problem if you farm without using chemicals. However, chemical poisoning can occur on farms subject to drift from aerial spraying or direct spraying on adjoining farms. If you have a stream or waterway running through your farm that originates on a farm where chemicals are used, you may have a concern. The chemicals may travel with the water in times of heavy rainfall or snowmelt, and your cow may be exposed. Poisoning may result from insecticides, pesticides, herbicides, or parasiticides. The best prevention is to be aware of these conditions and, if purchasing feed, being cautious about crops that may have been sprayed with chemicals prior to harvest.

Plant poisoning can occur if your pastures contain such toxic plants as bracken fern, nightshade, larkspur, hemlock, or others. Some poisonous plants, such as buttercup, grow in wet areas or along a ditch or stream. Plant poisoning is generally not a problem when enough grass is available in pastures. They are less palatable than grasses and will usually be avoided by cows. You should remove any that you find and develop a plan to prevent them from repopulating. A plant-poisoned animal is difficult to treat because the ailment is usually not discovered in time for treatment to be effective. Also, it may be difficult to determine which plant was eaten. The best prevention is to be aware of the poisonous plant populations in your region and survey your pastures for any that may be present and remove them.

Moldy feed toxicity occurs when a fungus or other mold grows in feed grains that are stored in moist conditions with poor ventilation. Usually the off-flavors or smells of such feeds will discourage your cow from eating them, but if she is hungry, she may eat it anyway. The easiest prevention is not to feed any grains, hay, or silage that has mold in it.

Acidosis occurs when the rumen develops an acidic condition with a pH of 4.0 to 4.5. This impairs rumen function and digestion, resulting in appetite loss. Gorging on large amounts of grain when she might have access to a feed cart or stored sacks of grain can cause acidosis. Avoid feeding her too much grain at the start of her lactation because her rumen bacteria populations have not increased enough to properly process these carbohydrates. During her dry period, these populations decrease because there is not enough grain to live on. They increase as more grain is added to the diet, but slowly at first. To treat acidosis, you will need to eliminate the grains and increase her fiber intake, such as dry hay. Also, magnesium carbonate and calcium hydroxide can be given as an oral drench to help neutralize the acidic rumen condition.

Hardware disease is not so much a disease or disorder as it is a condition from something your cow has swallowed, usually involving metal. Hardware disease results when a sharp object punctures the reticulum and the animal suddenly loses its appetite, has a reluctance to move, or moves gingerly. Wire is most often ingested if mixed in hay, silage, or grain. You can prevent hardware disease by orally giving a bolus-sized magnet to your cow once she arrives at your farm. This

A balling gun is a long metal tube with a hollow end for holding boluses or magnets. The gun is passed down the cow's throat and the ring plunger deposits the bolus or magnet below the cow's swallowing point. This prevents it from being coughed or spit out.

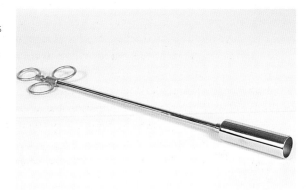

Aspirin and calcium boluses can be passed into the cow's stomach with a balling gun. Aspirin is sometimes used to ease a slight fever or to reduce the pain after an injury. Calcium boluses can be given at calving time to reduce the risk of milk fever.

Cow magnets are strong enough to pick up wire or metal that may be ingested. The metal sticks to the magnet and reduces the chance of puncturing the stomach lining or piercing the heart. The magnet rolls around the rumen for her entire life. It does not pass through the digestive system.

magnet will lie in her reticulum and move around in the churning material, and it will gather bits of metal and wire she may have swallowed. This will keep the hardware from puncturing her stomach wall or heart.

A **displaced abomasum** is when the cow's fourth stomach moves in the body cavity, twists, and prevents feedstuffs from passing to the digestive system. It is most often diagnosed by listening with a stethoscope and hearing a pinging sound when you tap the side of the cow below the loin ribs. Surgery is frequently required to relieve this problem. The operation involves suturing part of the stomach wall to the inside of the cow's body cavity to prevent a displacement from occurring again. Feeding bulky dry hay prior to and at calving time will help fill her stomach, lessening the possibility of the abomasum moving out of place.

Structural Disorders

Several physical problems in cattle originate because of structural conditions that may be exacerbated by environmental conditions. These problems include foot rot, hairy warts, and udder edema.

Foot rot is a condition due partly to physical structure of the foot and partly to environmental conditions. It is a break in the skin or hoof either between the toes or on the heel of the foot. When a break occurs, bacteria in dirt, manure, or the farmyard can enter these cracks and cause swelling and tenderness in the foot. The most common symptoms are progressive lameness and, in extreme cases, a swollen joint. Infections can move into the joint and spread into the leg or bloodstream. Before this happens, you will notice the tenderness and either treat it yourself or have your veterinarian help. A clean cow yard free of sharp objects that can break the skin or hoof will decrease the chance of infection. To treat a hoof infection, you will need to clean, disinfect, and wrap the hoof with a bandage.

A secondary condition affecting cow feet is **hairy warts**, which are lesions found on the heels of the feet. In later stages there are hairy-like protrusions

This is a healthy, well-formed hoof. It is rounded with short tips. This is important because it keeps an even pressure on both toes and avoids ankle injuries.

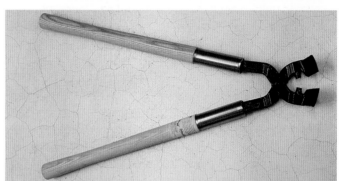

A cattle hoof trimmer is like a big toenail clipper. It is used to trim the tips of hooves that get too long. You need to restrain your cow if you trim her hooves or work with her feet.

Be cautious when trimming your cow's hooves. Use a hip restraint if you do it alone. Use short snips instead of large ones. Stop when you see a pinkish color in the white nail. Those are blood vessels and you shouldn't cut so deep that they start to bleed.

from the skin. Two or more microorganisms may work together to cause this condition. For the infection to start, the feet must be subjected to prolonged exposure to moisture and exclusion of air, typically found in muddy conditions. To prevent, keep your cow's stall dry and well-ventilated. Applying copper sulfate for several days has been found to help diminish the effects of this infection. The best prevention, however, is not buying a cow exhibiting hairy warts and bringing her onto your farm. Once these bacteria are introduced to your soil it is difficult, if not impossible, to eliminate them.

Udder edema is an excessive accumulation of fluid in the udder and usually occurs at calving time and shortly after. It tends to be more severe in high-producing or first-lactation cows. It takes four hundred pounds of blood pumped through your cow's udder to produce one pound of milk. Her heart has to work extremely hard pumping blood to maintain her health. If after calving, she can't move the swelling out quickly, she will be secreting milk at the same time, creating stress on her udder tissue. You can massage her udder and use diuretics to help relieve some of the symptoms, but generally these are only temporary effects. Time is the best healer for udder edema.

Reproductive Disorders

There are two conditions relating to reproduction that may affect the health of your cow, **metritis** and **endometritis**. These are inflammations of the uterus caused by bacteria, protozoa, fungi, or viruses that enter the cow's reproductive system at calving time. After calving, the uterus closes itself around the sphincter muscle. When closed, it retains any bacteria that may have entered at calving. This closed, warm, moist environment is an excellent incubator for bacteria, which usually cause infections. A clean calving pen or corral is the best preventative measure you can take, along with good hygiene if you need to assist a difficult birth. Treatment of these two conditions usually involves antibiotic infusions in the uterus.

A cow lying down may simply be resting. It's a pleasant sight. Your cow will get up to eat and lie down to chew her feed numerous times during the day. A cow that goes down due to foot injuries, calving paralysis, or milk fever is a different story and requires immediate attention.

Calves and Kids
Raising Up the Next Generation

■ ■ ■

DON'T KEEP all the fun to yourself. You can involve all members of your family in helping with the family cow; at least within reason and good sense. Many people choose to live on a farm to raise their children in the country rather than the city. Young people can get a good taste of farm life by helping with the work and feeding of a calf. Then they can graduate to an older cow. From the perspective of youngsters, she may appear huge. So big, in fact, that they may be intimidated the first time they see her.

But acquainting children with a small calf that is more their size may intrigue them. Calves are cute. They're lively and curious, and once trained to lead, they become docile and easy to handle. Start your youngster out with helping bottle-feed the newborn calf, if you don't let it nurse your cow.

Raising a Calf

After being encased in fluid for nine months, the newborn calf will try to stand up within the first hour. It will stagger and try to find the udder and teats of its mother. Watching a calf beginning to nurse is one of nature's wonders. It answers some silent call after it's born to know where to look for food. Most are successful in finding the udder, but some may need your help by directing them to it.

Some farmers remove the calf from the cow right away and strip or machine milk the colostrum from the udder for the calf's first meal. One reason for this practice is to train the calf to drink from a bottle right away. Immediate bottle-feeding eliminates the need to retrain the calf weeks later when it is removed from its mother. Some farmers allow the calf to nurse first to ensure that it gets colostrum milk quickly and then milk out the cow. Either method works.

If you let the calf nurse first and bottle-feed it later on, you can leave the calf and cow together. You should, however, monitor the calf to make sure it nurses. You may want to push the calf toward the cow's udder to get it into position to

Left: Let your child help take care of the calf. It is smaller and easier to be around than a large cow. Feeding is easy when using a bottle with a rubber nipple.

Working with animals is a wonderful experience for young family members. They learn responsibility, and it gives them a new awareness of how special rural life is.

eat. You can then direct its mouth towards the teat and encourage it. Most calves will catch on quickly once they've had a taste of the warm milk.

You should squeeze the teat and squirt some milk out to be sure it has an open end. Sometimes a hard, but small, gelatinous plug develops at the teat end that needs to be squirted out so the colostrum will freely flow. This plug is nature's way of protecting against bacteria growing through the teat end. Those few squirts are usually enough to signal milk let down to the cow's brain, and there will be plenty for the calf if it can master the process. That's one reason to monitor the calf's progress so that it receives the fullest benefits from that first milk.

Some calves can't figure out the mechanics of standing and sucking on a teat no matter how you try to help. This is not a real problem. It's probably frustrating for you although the hungry calf may feel it's more of a problem for it. You can turn this frustration into an opportunity by teaching it to drink from a bottle. This may have been your intent all along, and suddenly the calf is cooperating.

If the calf fails to drink from a bottle or nurse, you will have to provide some source of nutrition before it becomes dehydrated. Colostrum powders are available that can be mixed with water as a substitute for mother's milk. This will have to be prepurchased and on-hand for it to be useful.

If you plan to milk once a day and have the calf nurse the other time, you need the calf to learn right away. If it doesn't, you may need to milk twice a day until the calf gets the hang of it.

Bottle-feeding a calf is best done with a plastic bottle with a slip-on rubber nipple or one that has a plastic screw-on ring holding the rubber nipple in place. A calf naturally nurses with its head up. This is nature's way to close the esophagus leading to the lungs and open the trachea that leads to the stomach while gulping down milk. This valve closes as the head is tipped up and opens when the calf breathes between gulps of milk. This keeps the milk from getting into the lungs. The rubber slip-on nipple has a tiny air vent hole on the flat part of the nipple. You can poke the hole open with the tip of a small pocketknife to prevent the nipple from

Right: You may bottle-feed your calf or allow it to nurse directly from the cow. The approach you choose depends on the time you can invest and your own milk needs. *Shutterstock*

If you allow your calf to nurse from the cow, you will need to wean it at around six weeks of age. *Shutterstock*

collapsing by internal vacuum. This larger opening will allow more air to enter the bottle.

What if the calf is unwilling to suck on the nipple or the teat of its mother? You may have to use a nursing bag to feed it the first few times so that it gets the required nourishment. Plastic nursing bags can be bought at many local farm stores or ordered through farm supply catalogs specializing in dairy or beef products for feeding. They hold about a quart of milk, filled at the top. At the bottom of the bag is a long flexible tube attached at the end to a hard, hollow tube with a bulbous end. This is slipped down the calf's throat, past the esophagus opening to the lungs and into the trachea. When it is positioned correctly, the holding clip is opened and the milk drains into the calf's stomach until the bag is empty. Then the tube is slowly pulled out and the calf is fed.

It's essential that your calf get colostrum, the first milk produced by the cow. Colostrum contains antibodies and vitamins that will keep your calf healthy.

A nursing bag should only be used as a last resort for several reasons. First, it discourages the calf from learning how to suck to get its milk. You have delivered it without the calf doing any work. Second, the bulbous end of the tube can rip the fragile lining of the esophagus and cause irritation and possibly an infection. To prevent ripping, moisten the end of the tube with milk to make it slippery enough to slide down its throat easily. Finally, if not inserted correctly so that it bypasses the esophagus opening, it is possible to drain all the milk into the calf's lungs and drowning it instantly. There is no recovery when this happens.

Bottle-feeding is done by holding the bottle at an angle above the calf's head. This lets the milk drain down and decreases the likelihood of it getting into the lungs. If it doesn't get enough milk after getting a taste of it, the calf may butt or root around until it drinks some more. If a young member of the family feeds the calf, they need to be aware of this behavior so that they don't get hurt. But feeding a calf can be an excellent learning experience.

Feeding calves twice or three times a day is preferred to only once a day. A continual infusion of milk keeps the needed nutrition available. Depending on weather conditions, you may have to feed more often that three times a day. In extreme cold, it is better to feed smaller amounts more times per day than large quantities only a few times each day. This makes more work for you, but the calf needs it because it is burning more calories and fluids trying to keep itself warm than on moderate days. Even in cold weather, it should have access to clean, fresh water.

You can trade feeding whole milk for powdered milk replacers if you need your cow's milk for making other home dairy products. Powdered milk replacement products are available from most feed stores. Be sure to buy a high-fat milk replacer that is low in fiber. These products are more expensive, but high-fat products are more easily digested by calves than those with high fiber contents.

Typically, the colostrum period is finished in three to five days. Whole milk can continue to be fed to the calf until it's weaned at about six weeks of age. The transition from milk to solid food should be done over a period of one week. Increase the calf feed and decrease the milk given or dilute the milk with water to encourage the calf to eat the more specialized calf feed.

Calf starter feeds come in granular or pellet forms. They are high in protein, vitamins, and are easily digested. Because they often contain molasses, they are also tasty to the calf. A starter feed will be needed when you wean the calf from its mother because its digestive system will still be too immature to utilize grasses.

After a calf is weaned, it can be separated into another pen and fed alone. The mother or the calf (or both) may experience separation anxiety, but this only lasts for a few days. Some cows appear relieved they no longer have to look after a calf that has grown well.

A calf needs to feed at least two or three times per day. *Shutterstock*

In order to wean your calf, separate it from the mother and begin to feed it dry hay, grass, or calf starter feed.

The transition from milk to dry feed changes the calf's rumen characteristics. Although it still can utilize milk in its diet, other food sources, including dry hay, calf starter feed, and grass, will distract its attention away from its mother. You may develop a separate pen and pasture area for the calf because it will begin eating grass soon after weaning. It's a natural progression.

Calves die early most often because of three conditions: starvation, pneumonia, and scours and dehydration. Often these are cumulative effects with one leading to the other and compounding an already stressed physical condition.

Starvation is the result of not getting enough nutrition during the first few days. By not getting enough colostrum, the calf then lacks sufficient antibodies to protect against the cold, wet conditions that may exist. This can quickly lead to pneumonia. Pneumonia is often the result of manmade conditions. Too wet bedding, drafty pens, and poor ventilation all contribute to conditions that can lead to respiratory illnesses. Scours and dehydration need immediate intervention by providing electrolytes with fluids several times a day. At this point, the calf may or may not want to drink and you will be left with little choice but to feed with a plastic nursing bag. Sick calves don't like to eat any more than you do when you get sick. However, with a small calf, hours can make a difference between life and death. Fluids need to be given or the calf will continue to dehydrate.

Getting Your Children Involved

Where do young family members come in? With your guidance and initial help, your children can learn to feed and care for young calves. This duty provides good psychological development as well as physical activity. Being responsible for an animal outside the family pet assists their personal growth.

At that point, it is only a short step to your child developing a 4-H or FFA project. Activities offered through 4-H clubs and FFA chapters across the United States provide opportunities for young people to experience the pleasures and challenges of learning animal handling and raising.

Besides learning new skills while under experienced adult supervision and guidance, young members meet others their age and often with similar levels of experience and interest. This camaraderie can develop lifetime friendships. Both 4-H and FFA animal projects often are exhibited and shown at local and county fairs. While prizes, premiums, and recognition are part of the benefits of showing, even more importantly the chance of working with an animal over an extended period has many residual, often intangible, benefits for your kids.

It is easy for a child to become attached to a calf once she has started to care for it. An emotional connection begins and is carried through the life cycle from calf to cow.

Learning to feed, handle, and care for a calf are some of the positive experiences young people gain from participating in a county fair.

Aside from developing new friends with common interests, animal projects are excellent for developing life skills. Youth in 4-H and FFA develop record-keeping and budgeting skills. They are encouraged to assume total control for the animal's welfare. They are expected to keep records of the costs, expenses, income, and health of the animal in order to complete a project record at the end of the year.

Winning ribbons is part of the reward of showing good dairy animals. Developing personal skills, learning discipline, and valuing the ethical treatment of animals and fellow competitors are some of the most valuable rewards.

Other less tangible benefits will help your child's personal growth. Training an animal to lead by halter, which is required at shows and fairs, involves discipline, patience, and communication with an animal. The latter involves unspoken and verbal commands between your child and the animal and the use of hands on a halter to signal nonverbal direction. Your child's discipline involves daily dedication to animal care and scheduling the time to provide proper nutrition, water, and housing to ensure comfort and normal growth. Patience is required when training animals to lead by a halter. Steady, consistent commands are required because an animal cannot be expected to reason the same as a person.

By participating in shows, your child will learn about the ethics of raising and showing animals in livestock competition. Ethics are an important aspect of showing and help keep the competition fair for all exhibitors. Learning the ethics of properly handling and raising animals is key to the emotional and mental growth of a child. Understanding the ethical choices available and learning to make them is one of the most valuable lessons young people can learn.

4-H clubs and FFA chapters provide adult supervision to help children with an opportunity to develop leadership skills. Don't underestimate the importance of developing discipline, patience, and the ability to work with other members. These skills can provide a significant advantage as our children enter the job market.

People with experience in handling animals, ethics training, recordkeeping, leadership skills, and the ability to handle themselves in many different and challenging settings are always in demand by employers.

The rewards of doing a good job, competing fairly, being courteous to others, and being gracious in defeat are part of a dairy project experience.

Conclusion

Who would have thought that getting a family cow could lead to all these excellent opportunities? Milk, cheese-making, butter, composting, exercise, fencing skills, learning new ways to milk a cow, and youth projects are just a few of the things derived from her appearance on your farm.

Further opportunities will ripple out from your farm, especially for younger family members who learn about handling animals and about themselves. Character building comes from many sources. Your cow can be a big part of that and will be a big part of your new life.

Your new journey starts here, and it starts with your family cow.

Fresh milk and the dairy products that can be created from it are some of the rewards of keeping a family cow. *Shutterstock*

Opposite: Your cow will present you and your family with opportunities far beyond what you could have expected. *Shutterstock*

Acknowledgments

THERE ARE SEVERAL PEOPLE I'd like to thank for helping make this book possible. First, to my editor, Danielle Ibister, whose enthusiasm for and initiation of this topic presented me with the opportunity to write my fifth book for Voyageur Press. Her suggestions, comments, and critiques were always welcome. Her close attention to detail drew out many points that helped make this a better book.

To Dan and Paulette Johnson for their excellent photographic work that provides a visual dimension to the text. It was a pleasure to work with them as they willingly pursued the photographic subjects I requested.

To my wife, Mary, who shares my enthusiasm for dairy cattle and rural life, and whose suggestions, comments, and perspective always help focus my writing more clearly.

To our son, Marcus Hasheider, who filled in photos where needed and has now contributed his photographic work to a fifth book.

To Dan and Ellen Bohn, Daniel and Darlene Coehoorn, Charles Daggett, and Joel and Lisa Guell for use of their animals as photo subjects.

—P. H.

I'D LIKE TO SAY a big thank you to a lot of terrific folks who made the photography in this book possible. Doing the photography for a project like this takes a lot of time and effort and would be impossible without the friendly and helpful people I've met along the way. A particular thank you goes out to:

- Tom Mareth: Thanks, as always, for allowing me to photograph at your farm and for helping me get the photos I needed. It's really appreciated!
- Diane Rockafield: A good friend who'll cook for a book!
- Dale Carlson: I really enjoyed the morning spent at your farm. Thanks for sharing your cows and your knowledge.
- Dave and Jolynne Schroepfer: Your assistance with this project was greatly appreciated! Thanks for letting us photograph your cows and your farm. It was wonderful meeting you.
- The Langlade County Youth Fair exhibitors and dairy chairperson Mary Schmoll: A big thank you here! I had a great time at the fair and learned so much from so many devoted people.
- Danielle Ibister and everyone at Voyageur Press: I appreciated the opportunity to work on this project. Thank you!
- Paulette Johnson: As always, thank you for helping to coordinate photo shoots, for second-shooting, and for sharing your photo editing skills.
- Connie Summers: Thanks for providing a couple of images from your neck of the woods.

—D. J.

Resources

■   ■   ■

Ayrshire Breeders Association
1224 Alton Creek Road, Suite B
Columbus, OH 43228
(614) 335-0020
www.usayrshire.com

Brown Swiss Cattle Breeders Association of the USA
800 Pleasant Street
Beloit, WI 53511
(608) 365-4474
www.brownswissusa.com

American Dexter Cattle Association
4150 Merino Avenue
Watertown, MN 55388
(952) 215-2206
adca@dextercattle.org

Purebred Dexter Cattle Association of North America
25979 Highway EE
Prairie Home, MO 65068
(660) 841-9509
www.purebreddextercattle.org

Dutch Belted Cattle Association of America
c/o American Livestock Breeds Conservancy (ALBC)
P.O. Box 477
Pittsboro, NC 27312
(919) 542-5704
www.dutchbelted.com

American Guernsey Association
7614 Slate Ridge Boulevard
Reynoldsburg, OH 43068
(614) 864-2409
www.usguernsey.com

Holstein Association
1 Holstein Place
Brattleboro, VT 05302
(800) 952-5200
www.holsteinusa.com

American Jersey Cattle Association
6486 East Main Street
Reynoldsburg, OH 43068
(614) 861-3636
www.usjersey.com

American Lineback Dairy Cattle Association
c/o Elden Woolf, President
551 Northend Road
Mohrsville, PA 19541
(610) 916-1308
http://americanlinebacks.tripod.com

American Milking Devon Cattle Association
c/o Sue Randall, Registrar
135 Old Bay Road
New Durham, NH 03855
(603) 859-6611
www.milkingdevons.org

American Milking Shorthorn Society
800 Pleasant Street
Beloit, WI 53511
(608) 365-3332
www.milkingshorthorn.com

North American Normande Association
748 Enloe Road
Rewey, WI 53580
(608) 943-6091
www.normandeassociation.com

Red and White Dairy Cattle Association
3088 Ogden Avenue
Clinton, WI 53525
(608) 676-4900
www.redandwhitecattle.com

Glossary

abomasum. The fourth, true digestive stomach of a ruminant.

abortion. Loss of pregnancy before going full term.

artificial insemination (AI). Process in which a technician deposits frozen semen from a bull into the cow's uterus to create a pregnancy.

bagging. The rapid expansion of the udder in anticipation of calving, beginning up to two weeks beforehand.

balanced ration. A quantity of feed that fully satisfies a cow's daily requirements.

balling gun. Tool for placing a pill or bolus at the back of the mouth to make a cow or calf swallow it.

Bang's disease. Common name for brucellosis, a bacteria that causes abortion in cattle.

beef. The meat of a cow, bull, or steer.

beef type. Any cow showing a stocky, fleshier form that is favored by those concerned with meat production rather than an angular, refined form appropriate to dairy production.

bloat. Tight, swollen rumen caused by accumulation of gas; often occurs on left side when viewed from behind.

bolus. A large pill used for oral medications and treatments in animals.

bred. Was mated, or is considered pregnant.

bulk. The amount of physical space taken up by a food in relation to the nutrients it contains, often identified with hay and silages.

bull. Male bovine.

bulling. A slang term used to identify a cow in heat or estrus.

butterfat. Fat content in the milk that can be separated out to make cream or butter.

calving. Giving birth to a newborn calf.

cereal. Plants of the grass family that yield an edible, starchy grain, such as corn, wheat, oats, rye, barley, or rice.

cleanings. A slang term identifying the expelled placenta after calving.

clostridial disease. A deadly disease caused by spore-forming bacteria, including tetanus, blackleg, malignant edema, and entertoxemia.

Coccidiosis. Intestinal disease and diarrhea caused by protozoans.

colostrum. First milk after a cow calves, containing antibodies that give the calf temporary protection against certain diseases.

compost. A mixture of decomposed organic matter and used for fertilizer.

concentrate. A food high in energy or protein value, often lacking bulk.

conformation. General structure and shape of an animal.

cream line. The visible line that develops in time between cream and milk when whole, unhomogenized milk rests and the lighter-than-water butterfat globules rise to the top.

crossbreeding. The mating of a cow and a bull of different breeds.

cud. A wad of feed that a cow regurgitates from the rumen to be rechewed.

cycling. Having heat or estrus cycles.

dairy type. A cow having the body form preferred by those raising animals for milk, more angular than stocky.

dam. A female parent.

downer. A cow that is down on the ground and can't return to its feet due to any cause.

dry cow. Any cow that is not giving milk, usually in the two months prior to calving.

electrolytes. Chemical salts that ionize in solution and are helpful in treating dehydration as a result of scours; ions increase the electrical conductivity of solutions, often affecting muscle movements.

estrus. The period of ovulation and readiness for breeding, when the cow will accept the bull. Estrus repeats every nineteen to twenty-one days in sexually mature females.

flight zone. Distance you can get to an animal before it flees.

foot rot. Infection from soil bacteria in the hoof or ankle joint.

forage. Pasture and field crops that are of value to cattle for a diet of leaves and stems.

freemartin. A sexually imperfect female bovine that is sterile; this condition most often occurs when a female calf is born twin to a male.

freshen. To give birth to a calf and begin producing milk. A cow that has recently calved is referred to as a "fresh" cow.

grade. Any cow or bull not registered with one of the purebred associations.

grain. The seed of any member of the grass family, rich in easily digested starches.

grass. Plants with long leaves with parallel veins, jointed stems, and flowers and seeds at the ends of central stalks.

gutter. The shallow trench that collects the manure behind a cow's stanchion or tie stall.

hardware disease. A health problem caused by a sharp foreign object, such as accidentally swallowed wire penetrating the reticulum wall.

heat. A slang term for estrus.

heifer. A young female dairy cow. Usually used to describe a cow that has not yet had a calf; sometimes, however, a young cow that has calved may be called a "first-calf heifer."

hock. The flexible joint of the hind leg between the ankle and the hip, similar to the knee in humans.

hypocalcemia. Abnormally low blood calcium.

inbreeding. The breeding of closely related animals; it can take several forms, including sire to daughter, son to mother, or brother to sister. Inbreeding is used in an attempt to magnify good characteristics of an impressive individual but can also magnify the bad qualities.

intramuscular. Injection site into the muscle.

intravenous. Into the vein; often used to quickly infuse fluids into a cow's system.

ketosis. Incomplete metabolism of fatty acids, usually from carbohydrate deficiency or inadequate use of it; commonly seen in high-fat diets of cattle.

lactation. The physiological process of giving milk; it can also refer to a specific time period of milk production.

leaching. The downward percolation of minerals and organic nutrients through the soil.

legumes. Plants that are members of the trifoliate family because their leaves come in threes; include soybeans, peas, alfalfa, and clovers.

let down. Term used to identify the process of milk molecules being squeezed from the alveoli and released into the udder canals, and ready for removal through the teat structures.

linebreeding. A mild form of inbreeding in which offspring are bred back to one common ancestor.

mastitis. An infection of the mammary gland caused by bacteria.

maternal traits. Characteristics of a good mother cow.

metritis. Inflammation of the uterus.

mycotoxin. Toxic substance found in moldy hay, grain, or silage.

omasum. The third stomach of a ruminant located between the reticulum and abomasums.

open. A cow or heifer of breeding age that is not pregnant or has not been bred.

pasteurize. To heat milk to a certain temperature to kill bacteria.

peritonitis. Inflammation of the lining of the abdominal cavity; usually results from a puncture wound or hardware that pierces the stomach wall and fermentation fluids enter the space between the body wall and internal organs.

PI3. Parainfluenza 3, a viral disease of cattle that can cause respiratory problems.

pin bone. The posterior ends of the pelvic bones in cattle. They appear as two raised areas on either side of the tail head.

pinkeye. Contagious eye infection spread by flies.

pneumonia. Infection of the lungs.

polled. A genetically hornless cow.

production records. Measure of milk produced in a certain time period.

purebred. A cow born of parents belonging to a nationally recognized breed or that has been registered in one of the national association herd books.

quarter. One of four compartments of the cow's udder.

ration. The daily diet of a cow, formulated to provide all required daily nutrients.

raw milk. Non-pasteurized milk straight from the cow.

registered. A cow certified to be purebred through the registration with one of the breed associations.

retained placenta. A placenta that fails to shed after calving.

reticulum. The second stomach of a ruminant. It has a honeycomb structure of hexagonal cells.

ringworm. Fungal infection causing scaly patches of skin; highly contagious.

roughage. A feed that is high in bulk and fiber and low in digestible nutrients.

rumen. The large first stomach compartment of a ruminant.

scours. Diarrhea, used mainly in reference to calves.

settle. A term used to identify an animal that conceived from breeding; failing to conceive is referred to as "not settling."

sire. A male parent.

silage. A roughage that is preserved through fermentation, such as corn, oats, wheat, alfalfa, or grass. It is most often stored in an upright silo, plastic bags, trenches, or cement bunkers.

skimmed milk. Milk from which most of the butterfat has been removed.

solids. The non-water ingredients of milk that include fat, proteins, and minerals.

solids-non-fat. The total solids in milk without the butterfat protein.

springer. A cow or heifer that is about to calve.

stanchion. A long wooden or metal upright frame that clamps loosely behind a cow's head, confining her to a stall while allowing her forward and backward movement and allowing her to lie down.

standing heat. A period of time during estrus when a cow will stand still while being mounted by a bull, or another cow.

steer. A castrated male bovine.

strip. To hand-milk a cow. To *strip out* is to squeeze all of the milk out of the udder down to the last drops.

supplement. Any addition to a balanced ration that may be needed; often vitamins or minerals.

tattoo. Permanent mark in the ear often signifying vaccination for brucellosis.

teat. The elongated protuberances on a cow's udder where the milk exits; pronounced *teet*.

thurl. The hip joint in cattle located on either side of the pelvis.

trocar. A sharp-pointed instrument equipped with a cannula, used to puncture the wall of a body cavity to release gas or withdraw fluid; most often used in cases of bloat.

udder. The body of the cow's mammary gland divided internally into four separate quarters, each equipped with one teat.

vaccine. A fluid containing killed or modified germs, put into an animal's body to stimulate the production of antibodies and immunity.

warble. Larva of the heel fly; it burrows out through the skin on the cow's back.

waste milk. Milk that can be used for calves, pigs, sheep, or goats but not for human food; often associated with colostrum or milk containing antibiotics after treating for mastitis.

wean. To separate a calf from its mother or stop feeding the calf milk.

whey. The watery part of milk that is separated from the coagulable part or curd, especially in the process of cheesemaking.

withdrawal time. The interval between drug administration and the time of legal slaughter for the sale of meat or milk; also refers to the length of time it takes for a drug to disappear from the system of the treated animal.

yearling. Refers to a calf usually between one and two years of age.

Index

A

acidosis, 191–192
agricultural extension service, 22
America Lineback, 47, 60
antibiotics, 123, 145, 175–179, 195
aspirin, 192
auction barn sales, 28–31, 45
Aussie Red, 57
Ayrshire, 47, 48, 57

B

balling gun, 192
Barnum, P. T., 56
bedding, 65–68, 70, 71, 75, 97, 99, 104, 105, 138, 149, 184, 203
bloat, 182–184
body temperature, 173–174
  rectal thermometer, 174
breeding, 11, 20, 42, 43, 45, 129, 130, 138–143
  artificial insemination (AI), 129, 141–143
  natural service, 141, 143
Brown Swiss, 47, 49, 50
brucellosis (Bang's disease), 26–27, 186–187
butter-making, 9, 22, 55, 108, 147, 151–153
  electric butter churn, 151
  for kids, 154–155
buttermilk, 151–152, 155, 164, 167

C

calcium boluses, 192
calves
  bottle feeding, 117, 197, 199, 201
  dehydration, 203
  involving children, 204–207
  nursing, 197–201, 203
    bags, 201
  raising, 197–203
calving, 13, 30, 36, 44, 129–145
  how to handle problems, 133, 136
  paralysis, 137–138
  positions, 134–135
  what to do, 130–132
  what to watch for during and after, 136–138

cattle hoof trimmer, 194
changes in American farming and dairying practices, 11–12
cheese-making, 9, 19, 22, 108, 147, 161–171
    aging, 162
    cautions, 167
    cheddar, 160, 162
    cottage cheese, 160, 170–171
    equipment, 162, 163
        cheese press, 163, 168, 169
    ingredients, 160, 163
        rennet, 163–165
        salt, 166, 169
        starter, 163–165, 167
    Swiss, 167
choosing a cow, 9, 25–45
    picking a breed, 48–61
        crossbreds, 47, 57–60
        purebreds, 43, 48–57
    what to look for, 34–41
clostridial diseases, 184
conformation, 35–38
consignment sales, 30
cow magnets, 193
cream, 150, 151

D

dairy soap, 118–119
displaced abomasums, 193
Devon, 55, 59
Dexter, 47, 54, 55, 140
Dutch Belted, 47, 55–56
Dutch Friesian, 50

E

endometritis, 195

F

fatty liver syndrome, 189–190
feeding your cow, 81–105
    grains, 88
    hay, 81, 86–88, 92–95
    how much, 89–91
    how to, 85
    minerals, 88–89
    purchasing feed, 86
    vitamins, 89
fencing, 63, 76
    gates, 63
    temporary versus permanent, 76–77
    wire, 78–79
        electric, 79
FFA, 204, 206–207
foot rot, 193

4-H, 204, 206–207
front-end loader, 101

G

giving your cow access to water, 71, 73
goat milk, 160
grass tetany, 189–191
grazing networks, 22–23
Great Depression, 11
Guernsey, 44, 49–50

I

ice cream, 9, 55, 147, 159
        recipe, 159
illness
        signs of, 173, 177
incubator, 156–157

H

hanging scale, 90
hardware disease, 192
hay baler, 87–88, 95
hip restraint, 124, 194
Holstein, 10, 25, 47, 48, 50–51, 54, 57, 89, 140
        Holstein-Jersey cross, 57
homeopathic and herbal treatments, 179–181
horns, 42
housing your cow, 63–75
        hutches or huts, 74
        pens or stalls, 70–71, 75, 76
        planning simple structures, 67–69

J

Jersey, 47, 51–52, 57, 89, 162. See also Holstein-Jersey cross.
Johne's disease, 45

K

Kerry, 55
ketosis, 189–190

L

lactation, 29, 31, 35–36, 41, 45, 88, 107, 115–117, 129, 141
        colostrum, 123, 145, 184, 197, 202
        dry period, 44, 81, 107, 110, 132, 140, 144, 145
        let-down, 117, 118, 122, 199
        stage of, 32, 35, 44, 113
livestock ordinances, 16

M

manure, 18, 44–45, 97–105
        cart, 102
        composting, 17–18, 97, 100–103
            bins, 103

pack, 65
pats, 98–99
removing, 68–70
waste runoff, 97, 102, 103
metritis, 195
milk fever, 137–138, 189
milk production, 10, 20, 25–26, 41, 47, 88, 90–91, 107, 140, 162
adjusting, 108, 110, 144–145
by breed, 49–60
records, 127
milking
hand-milking, 38, 39, 41, 48, 72, 110, 113–117
how to, 117–121
machines, 12, 38, 113–115, 117
using, 122
pail, 122
safe practices, 124
setting a schedule, 108, 110–111
teaching a cow how to, 126
Milking Devon, 47, 59
Milking Shorthorn, 47, 53, 57
moldy feed, 94–95, 191

N

Normande, 47, 58
Norwegian Red, 57

O

older versus younger cows, 43

P

parasites, 185
pasteurization, 147, 149, 156, 162, 164
pasture, 17, 81, 94, 95
calculating how much your cow needs, 17–18, 91–92
grasses, 84
pinkeye, 188
pneumonia, 184, 203
poisoning, 189, 191
plants, 94
pregnancy, 26, 27, 29, 31, 129, 130, 137–139, 140–142, 144, 186
protecting your cow from heat and cold, 63–65
windbreak, 66–67

R

raw milk, 81, 147, 149, 150, 156
straining, 147, 149
selling of, 20
Red and White Dairy Cattle Association (RWCA), 54
Red and Whites, 47, 54
Red Angler, 57
Ringworm, 188
ruminant digestive system, 81–83

abomasum, 82–83
omasum, 82–83
reticulum, 82–83
rumen, 82, 83

S

scours, 184, 203
self-sufficient lifestyle, 15–16
sheep milk, 160
Shorthorn, 48, 53
skid-steer loader, 69, 88, 100, 104
spreader, 105
stanchion, 72

T

tie-stall, 72
tuberculosis (TB), 26–27, 147, 187–188

U

udder, 38–43, 45, 52
    edema, 195
    expansion of, 130
    keeping it healthy, 115–117
    mastitis, 39, 116, 127, 177, 178
    sanitizing, 118–119, 123
United States Department of Agriculture (USDA), 23
uterus
    torsion of the, 136–137
    uterine prolapse, 137

V

vaccination, 175, 184, 186
ventilation, 74–75
veterinary assistance, 175, 183

W

warts, 189
    hairy, 193–194
water buffalo milk, 160
World War I, 11
World War II, 11

Y

yogurt, 9, 147, 156–157

Right: A solitary cow under a rainbow presents an idyllic rural scene.

About the
Author and Photographer

■ ■ ■

Courtesy of Marcus Hasheider

PHILIP HASHEIDER grew up on a Wisconsin dairy farm where he was involved in 4-H and FFA projects. He is the author of twelve books, including *How to Raise Cattle*, *How to Raise Pigs*, *How to Raise Sheep*, and, most recently, *The Complete Book of Butchering: Smoking, Curing, and Sausage Making*.

Philip is the 2005 recipient of the Book of Merit award presented by the Wisconsin Historical Society and Wisconsin State Genealogical Society. He has written numerous articles for national and international dairy breed publications. His diverse work has appeared in the *Wisconsin Academy of Review*, *The Capital Times*, *Wisconsin State Journal*, *Sickle & Sheaf*, *Old Sauk Trails*, *Sauk Prairie Area Historical Society Newsletter*, *Sauk Prairie Eagle*, and *Holstein World*. He was the writer for the *Wisconsin Local Food Marketing Guide* for the Wisconsin Department of Agriculture, Trade, and Consumer Protection. His essays have appeared in two collections, *Seasons on the Farm* and *My First Tractor*.

A Dairy Science degree from the University of Wisconsin provided the background to help Philip manage his family's dairy for twenty years. He also has seven years experience as a cheesemaker's assistant. Philip works as a Dairy Cattle Analyzer with aAa® Animal Analysis, a private international organization working with all dairy cattle breeds. He lives with his wife and two children on their farm near Sauk City, Wisconsin.

Courtesy of Paulette Johnson

DANIEL JOHNSON is a full-time professional photographer and writer specializing in farm imagery. He helps run Fox Hill Farm, a family-owned horse farm in far northern Wisconsin. A 4-H alumnus, he is the author and photographer of the *4-H Guide to Digital Photography*. His images are featured in numerous books, magazines, calendars, and greeting cards. Dan's work can be viewed at www.foxhillphoto.com or on the Fox Hill Farm blog at www.foxhillphoto.blogspot.com. He lives in far northern Wisconsin, near the town of Phelps.